T0193814

Metasurface Holography

Synthesis Lectures on Materials and Optics

© Springer Nature Switzerland AG 2022

Reprint of original edition © Morgan & Claypool 2020

All rights reserved. No part of this publication may be reproduced, stored in a retrieval system, or transmitted in any form or by any means—electronic, mechanical, photocopy, recording, or any other except for brief quotations in printed reviews, without the prior permission of the publisher.

Metasurface Holography

Zi-Lan Deng, Xiangping Li, and Guixin Li

ISBN: 978-3-031-01258-7   paperback
ISBN: 978-3-031-02386-6   ebook
ISBN: 978-3-031-00250-2   hardcover

DOI 10.1007/978-3-031-02386-6

A Publication in the Springer series

SYNTHESIS LECTURES ON MATERIALS AND OPTICS

Lecture #4
Series ISSN
Synthesis Lectures on Materials and Optics
Print 2691-1930   Electronic 2691-1949

Metasurface Holography

Zi-Lan Deng
Jinan University, Guangzhou, China

Xiangping Li
Jinan University, Guangzhou, China

Guixin Li
Southern University of Science and Technology, Shenzhen, China

SYNTHESIS LECTURES ON MATERIALS AND OPTICS #4

ABSTRACT

The merging of metasurface and holography brings about unprecedented opportunities for versatile manipulation of light in terms of both far-field wavefront and near-field profile. In this book, a brief evolving history from surface plasmon polariton holography to metamaterial holography and finally to metasurface holography is introduced at first. Basic physical mechanisms that govern the phase modulation rules behind metasurface holography design are discussed later. Next, extended functionalities such as arbitrary polarization holography, vectorial holography, full-color holography, and hybrid holography achieved in the metasurface platform are presented. Surface wave and metagrating holography that bridges the on-chip surface wave and free-space wave is also introduced. In the end, we envisage practical applications of high-fidelity 3D holographic display, high-secure encryption, and high-capacity digital encoding and also indicate remaining challenges based on metasurface holography.

KEYWORDS

metasurface holography, holography design, phase modulation, 3D holographic display, high-secure encryption, digital encoding

Contents

CHAPTER 1

Introduction and Outline

Manipulation of light fields at will is a long-sought goal in the optics community. In this regard, holography, a technology that can modulate an arbitrary light wavefront including not only its intensity but also the phase information in a spatial domain, was invented 60 years ago [1]. Traditional holography has been extensively studied by the information optics community and has made tremendous progress in the past years, leading to many practical applications such as holographic storage [2], interferometry [3], lithography [4], and three-dimensional (3D) displays [5].

On the other hand, in the emerging nanophotonics field, metasurface, composed of arrays of two-dimensional (2D) subwavelength structures, represents a much more powerful and compact platform for manipulation of light field [6]. Beyond the amplitude and phase information of light field, many other aspects of light including its polarization, frequency, momentum, and angular momentum, can be fully controlled with the metasurface platform.

The fields of holography and nanophotonics were perfectly combined with each other for a long history. It could be retrospected to Cowan's seminal work using surface plsmon polariton (SPP) as a reference beam in 1972 [7]. Kawata et al. once employed SPP to realize evanescent-wave holography [8, 9] and later for the full-color holography [10]. Afterward, many surface plasmon holography devices based on plasmonic nanostructured grating structures have been proposed to combine the surface plasmon wave and the free-space wave [11]. The gradient phase array that forms the basis of metasurface was first proposed by Lalanne et al. in 1998 [12]. And the report of arbitrary wavefront shaping metasurfaces based on the such phase array concept was later proposed by Hasman in 2002 [13] and Capasso et al. in 2011 [6], respectively. From then on, metasurface holography has been extensively studied toward various types of applications typically including holographic imaging [14], angular momentum generation [15], light accelerating [16], efficient coupling between surface and freespace wavefront [17], and so on.

The underline physical mechanism for holography is controlling the spatially varying phase retardation profile in the cross section of a light beam. Conventionally, its physical realization relies on the generation of spatially varying accumulated phase retardation through traditional diffractive optical element (DOE). The accumulated phase originates from the undergone light path in the traditional diffractive optical element (DOE) with fixed refractive index. Thickness of traditional DOE is controlled to modulate the spatially varying phase profile. With the metasurface platform, however, the wavefront shaping mechanism is totally different from the traditional DOE. It relies on the abrupt phase discontinuity on an ultrathin planar interface (subwavelength or wavelength scale) by engineering the scattering properties of the individual building blocks (meta-atoms) of the metasurface in a spatially varying way. Usually, the shape, size, orientation, and displacement of meta-atoms composed of plasmonic nano-antennas or all-dielectric meta-atoms were tailored on demand to realize the given wavefront transfer functionalities. The overview of the evolution process of metasurface holography will be described in Chapter 2. From the point of view of underlying physical mechanism, the typical modulated phase in metasurfaces can be classified as resonant phase [18], geometric phase [19], propagation phase [20], and detour phase [21], which will be summarized in Chapter 3. Different phase types have different characteristics and can be combined together to build multifunctional devices within a single lightweight interface, which is hardly feasible by the traditional holography technology. Beyond the tradition holography that mainly deal with phase and amplitude of light, polarization states can also simultaneously be controlled at will in the metasurface platform. A variety of polarization multiplexed holograms will be included in Chapter 4. In addition, frequency of light can also be well manipulated in the metasurface holography design, which will lead to color-controlled holography and full-color holographic imaging, as discussed in Chapter 5. As a unique property possessed by metasurface holography, the direct shaping of either surface wavefront or free-space wavefront at a metagrating will be summarized in Chapter 6.

In brief, metasurface holography provides much more degrees of freedom for light field manipulation than traditional holography technology. It can integrate multiple functionalities with high performances in a single ultrathin layer. Although there are still some challenges, such as the dynamic tunability issue and high cost fabrication, metasurface holography still represents a big step toward versatile manipulation of light fields at will.

1.1   REFERENCES

[1] D. Gabor, A new microscopic principle, *Nature*, 161:777–778, 1948. DOI: 10.1038/161777a0. 1

[2] J. F. Heanue, M. C. Bashaw, and L. Hesselink, Volume holographic storage and retrieval of digital data, *Science*, 265(5173):749, 1994. DOI: 10.1126/science.265.5173.749. 1

[3] L. Heflinger, R. Wuerker, and R. E. Brooks, Holographic interferometry, *Journal of Applied Physics*, 37(2):642–649, 1966. DOI: 10.1063/1.1708231. 1

[4] M. Campbell, D. Sharp, M. Harrison, R. Denning, and A. Turberfield, Fabrication of photonic crystals for the visible spectrum by holographic lithography, *Nature*, 404(6773):53, 2000. DOI: 10.1038/35003523. 1

[5] S. Tay, P.-A. Blanche, R. Voorakaranam, A. Tunç, W. Lin, S. Rokutanda, T. Gu, D. Flores, P. Wang, and G. Li, An updatable holographic three-dimensional display, *Nature*, 451(7179):694, 2008. DOI: 10.1038/nature06596. 1

[6] N. Yu, P. Genevet, M. A. Kats, F. Aieta, J.-P. Tetienne, F. Capasso, and Z. Gaburro, Light propagation with phase discontinuities: Generalized laws of reflection and refraction, *Science*, 334(6054):333–337, 2011. DOI: 10.1126/science.1210713. 1

[7] J. J. Cowan, The surface plasmon resonance effect in holography, *Opt. Commun.*, 5(2):69–72, 1972. DOI: 10.1016/0030-4018(72)90001-6. 1

[8] S. Maruo, O. Nakamura, and S. Kawata, Evanescent-wave holography by use of surface-plasmon resonance, *Appl. Opt.*, 36(11):2343–2346, 1997. DOI: 10.1364/ao.36.002343. 1

[9] G. P. Wang, T. Sugiura, and S. Kawata, Holography with surface-plasmon-coupled waveguide modes, *Appl. Opt.*, 40(22):3649–3653, 2001. DOI: 10.1364/ao.40.003649. 1

[10] M. Ozaki, J.-i. Kato, and S. Kawata, Surface-plasmon holography with white-light illumination, *Science*, 332(6026):218–220, 2011. DOI: 10.1126/science.1201045. 1

[11] I. Dolev, I. Epstein, and A. Arie, Surface-plasmon holographic beam shaping, *Phys. Rev. Lett.*, 109(20):203903, 2012. DOI: 10.1103/physrevlett.109.203903. 1

[12] P. Lalanne, S. Astilean, P. Chavel, E. Cambril, and H. Launois, Blazed binary subwavelength gratings with efficiencies larger than those of conventional échelette gratings, *Opt. Lett.*, 23(14):1081–1083, 1998. DOI: 10.1364/ol.23.001081. 1

[13] Z. E. Bomzon, G. Biener, V. Kleiner, and E. Hasman, Space-variant Pancharatnam–Berry phase optical elements with computer-generated subwavelength gratings, *Opt. Lett.*, 27(13):1141–1143, 2002. DOI: 10.1364/ol.27.001141. 1

[14] G. Zheng, H. Mühlenbernd, M. Kenney, G. Li, T. Zentgraf, and S. Zhang, Metasurface holograms reaching: 80% efficiency, *Nat. Nanotechnol.*, 10(4):308–312, 2015. DOI: 10.1038/nnano.2015.2. 1

[15] F. Yue, D. Wen, C. Zhang, B. D. Gerardot, W. Wang, S. Zhang, and X. Chen, Multi-channel polarization-controllable superpositions of orbital angular momentum states, *Adv. Mater.*, 29(15):1603838–n/a, 2017. DOI: 10.1002/adma.201603838. 1

[16] M. Henstridge, C. Pfeiffer, D. Wang, A. Boltasseva, V. M. Shalaev, A. Grbic, and R. Merlin, Accelerating light with metasurfaces, *Optica*, 5(6):678–681, 2018. DOI: 10.1364/optica.5.000678. 1

[17] S. Sun, Q. He, S. Xiao, Q. Xu, X. Li, and L. Zhou, Gradient-index meta-surfaces as a bridge linking propagating waves and surface waves, *Nat. Mater.*, 11(5):426–431, 2012. DOI: 10.1038/nmat3292. 1

[18] F. Zhou, Y. Liu, and W. Cai, Plasmonic holographic imaging with V-shaped nanoantenna array, *Opt. Express*, 21(4):4348–4354, 2013. DOI: 10.1364/oe.21.004348. 2

[19] M. Kang, T. Feng, H.-T. Wang, and J. Li, Wave front engineering from an array of thin aperture antennas, *Opt. Express*, 20(14):15882–15890, 2012. DOI: 10.1364/oe.20.015882. 2

[20] A. Arbabi, Y. Horie, A. J. Ball, M. Bagheri, and A. Faraon, Subwavelength-thick lenses with high numerical apertures and large efficiency based on high-contrast transmitarrays, *Nat. Commun.*, 6:7069, 2015. DOI: 10.1038/ncomms8069. 2

[21] Z.-L. Deng, J. Deng, X. Zhuang, S. Wang, T. Shi, G. P. Wang, Y. Wang, J. Xu, Y. Cao, X. Wang, X. Cheng, G. Li, and X. Li, Facile metagrating holograms with broadband and extreme angle tolerance, *Light Sci. and Appl.*, 7(1):78, 2018. DOI: 10.1038/s41377-018-0075-0. 2

CHAPTER 2

Overview of the Emerging Field of Metasurface Holography

The holography technology was invented by Dennis Gabor in 1948 when he tried to correct the spherical aberrations of the electron microscopy by introducing a so-called "electron interference microscope" [1]. The proposed principle provided a complete record of amplitudes and phases in one diagram, and since then was widely studied to reconstruct both 2D and 3D images in the optics community.

Hologram, which records the spatial amplitude and phase profile in a 2D plane, plays a central role in the holography technology. In earlier times, holograms were produced by complex optical setups that create the interference between a reference wave (usually a plane wave) and an object wave that is to be recorded. Afterward, computer-generated holograms (CGHs) [2, 3] that record the object information using computational algorithm becomes more popular, as there is no need for a real interference between two waves, and nonexistent virtual 3D objects can also be recorded by the CGHs. The CGHs can be flexibly implemented in dielectric materials with spatially varying thickness or the liquid crystal-based spatial light modulators (SLMs), which modulate the wavefront by the accumulated propagation phase. Recently, the traditional holographic technology was combined with the emerging nanophotoics field through surface plasmon polaritons [4] and metamaterials [5]. Later on, novel phase modulation rules were introduced by single-layered metasurfaces [6] and these structured ultra-thin surfaces emerged as a powerful platform to demonstrate multifunctional and high-performance holographic applications.

2.1   BASIC WAYS TO CONSTRUCT HOLOGRAPHY

The basic procedures to construct holography are the same for both traditional holography and nanophotonic holography, expect that the physical implementations of the hologram are different. Holographic plate was a widely used medium for hologram recording. The spatially modulated phase profile in the holographic plate is based on the accumulated phase change when the light pass through the constituent emulsion or photo-reduced polymer. The physical carrier of the CGH was usually implemented by SLM, which modulate the light wavefront by controlling the propagation phase of light passing through the anisotropic liquid crystal elements. Due to the electric tunablity of the liquid crystal, the SLM enables the dynamical holographic imaging and displays. However, the pitch size of the SLM is as large as tens of micrometers, significantly limiting the view angles ($\sim 4°$) and diffraction efficiencies with multiple unwanted high diffraction orders [7, 8]. In the metasurface community, the phase modulation rules are fundamentally different from the traditional ways, and all attributes of light including its amplitude, phase, polarization, and frequency can be controlled within the subwavelength scale, which not only improves the performance but also extends the functionalities of holographic applications.

Even if the phase modulation rules and physical carriers for hologram are completely different in traditional holography and metasurface holography platforms, they share the same algorithm to build the hologram from a given 2D or 3D object wavefront. In general, the wavefront of an object wave is described by the spatial profile of its complex-amplitude $I(x_i, y_i)$ at the image plane. After propagating a distance z_d, the complex-amplitude information of the wavefront is recorded as the hologram $H(x_h, y_h)$. The relation between the complex-amplitude profiles of the image and hologram is dictated by the Huygens-Fresnel principle [9],

$$H(x_h, y_h) = \frac{1}{j\lambda} \iint I(x_i, y_i) K(\theta) \frac{\exp(ikr)}{r} dx_i dy_i, \tag{2.1}$$

where (x_h, y_h) and (x_i, y_i) are the coordinates of hologram and image frames, respectively, λ is the wavelength of light, $r = \sqrt{(x_h - x_i)^2 + (y_h - y_i)^2 + z_d^2}$, and $K(\theta)$ is an obliquity factor. Directly calculating the integration of Eq. (2.1) is unpractical due to the uncertainty of the obliquity factor. For the numerical implementation, the following Fresnel and Fraunhofer approximations applicable for different diffraction conditions were extensively applied.

The Fresnel diffraction is applicable when the distance between the hologram and image z_d is much larger than the feature sizes of both the hologram and image, where the paraxial approximation can be used to simplify the diffraction formula as follows:

$$H(x_h, y_h) = \frac{\exp(j2\pi z_d/\lambda)}{j\lambda z_d} \exp\left[\frac{j\pi(x_h^2 + y_h^2)}{\lambda z_d}\right] F\left[I(x_i, y_i) \exp\left[\frac{j\pi(x_i^2 + y_i^2)}{\lambda z_d}\right]\right], \tag{2.2}$$

where $F[\cdot]$ stands for the Fourier transformation (FT) evaluated using the spatial frequencies $(x_h/\lambda z_d, y_h/\lambda z_d)$. In the practical calculation, the mature Fast Fourier transform (FFT) algorithm is used to significantly speed up the computation. Compared with the direct integration method, FFT can reduce the computation complexity from $O(N^4)$ to $O(N \log N)$, where N is the resolution of the hologram.

The Fraunhofer diffraction is applicable at a more stringent condition, $z_d > \frac{\pi}{\lambda}(x_i^2 + y_i^2)_{max}$, which further simplifies the diffraction formula as follows:

$$H(x_h, y_h) = \frac{\exp(j2\pi z_d/\lambda)}{j\lambda z_d} \exp\left[\frac{j\pi(x_h^2 + y_h^2)}{\lambda z_d}\right] F[I(x_i, y_i)]. \tag{2.3}$$

From the Fraunhofer formula, one can directly obtain the hologram from the FT of the image profile. Thus, the holograms calculated from Fraunhofer formula are also referred to as Fourier holograms. In the practical situation, people typically place a lens between the hologram plane and the image plane, so that the effective distance between the image plane and the hologram plane can satisfy the stringent condition for Fraunhofer diffraction. In the hologram calculation of a 2D image, both the Fresnel and Fraunhofer formulas were usually combined with the Gerchberg and Saxton's iterative algorithm [10] to finally produce a phase-only hologram that was easily to implement in physical medium.

To compute the hologram of a 3D object, one can discretize the 3D object into a sets of points or polygons. The wavefront generated by the 3D object can be described by the superposition of all the spherical waves from all discretized points or polygons,

$$H\left(x_h, y_h\right) = \frac{1}{j\lambda} \sum I\left(x_i, y_i, z_i\right) \frac{\exp\left(ikr_i\right)}{r_i}. \tag{2.4}$$

To speed up the computation of the above summation, FFT algorithm can be applied to replace partial steps in the summation operation of the above computation [8, 11].

2.2   EVOLUTION FROM TRADITIONAL HOLOGRAPHY TO METASURFACE HOLOGRAPHY

The implementation of holography in nanophotonics platform can be traced back to 1972 [12], followed by a few reports that use SPP to enhance the diffraction efficiency [13, 14]. In 2011, Kawata et al. applied the SPP to 3D color holography with white-light illumination, which opened up a novel route for SPP holography [4]. The hologram in their configuration was still recorded on a photoresist through traditional interference method. In addition, they coated a thin metal on the photoresist hologram to excite the SPP waves. Based on the dispersion relation of the SPP, only the wavelength that satisfies the resonance conditions of the SPP mode can be reconstructed in the diffraction space. In this way, by combing three white-light beams with three different incident angles, only three wavelengths can be filtered out to take part in the holographic reconstruction, and different wavelengths can carry different holographic information. In the observation zone, the diffracted light waves from the three wavelengths exhibit independent images coming from the multiplexed holograms, and finally a full-color 3D holographic image can be constructed [4].

The above SPP holography still records the hologram in traditional way, and SPP plays an assistant role for the wavelength filtering. In 2012, the multilayered metamaterial holography was proposed by modulating the phase with spatially varying effective medium [15]. Metamaterial is a kind of artificial structure composed of an array of subwavelength resonant elements, capable of designing arbitrary effective medium that exhibits unnatural effective refractive index, i.e., negative index [16], zero index [17], and extremely high index [18]. Previously, metamaterials were mainly used to mold light propagation inside the structure, where the near-field

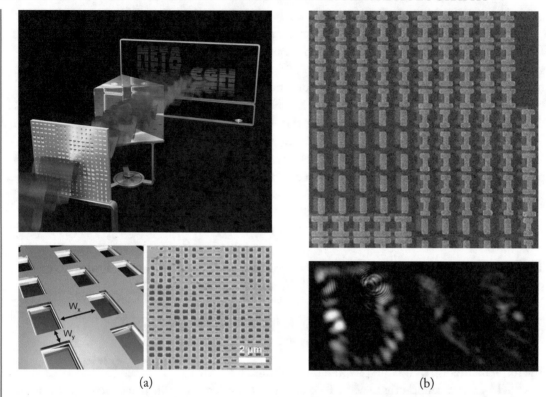

(a)   (b)

Figure 2.1: Metamaterial holography that shapes the wavefront by accumulated phase retardation when the light goes through the effective medium formed by bulky metamaterials. (a) A multi-wavelength holography implemented by a fishnet metamaterial [15]. (b) Infrared phase holograms by a multilayered I-shaped metamaterial [5].

of light is controlled on demand by tailoring the effective permittivity and permeability of the metamaterial based on the framework of transformation optics [19]. Instead, Fig. 2.1 shows the works that employ metamaterials to shape far field light in the free space. In Fig. 2.1a, a fishnet metamaterial was exploited to realize spatial and spectral light shaping. The size and geometry of the unit cell structure were properly tailored to have two different transmittances for each of the two predesigned wavelengths. Two binary holograms are modulated by those two sets of unit cell structures, and eventually the metamaterial emits two different letters: "META" and "CGH," for two different wavelengths. Figure 2.1b shows another metamaterial holography with a multilayered metal-bar arrays. The size and geometry of the metal-bar were tailored for a certain range such that the effective refractive index varied in a range from 2–5. As a result, the accumulated propagation phase after the incident light passes through the metamaterial can be tailored in a spatially varying way. In this way, an infrared metamaterial phase hologram was

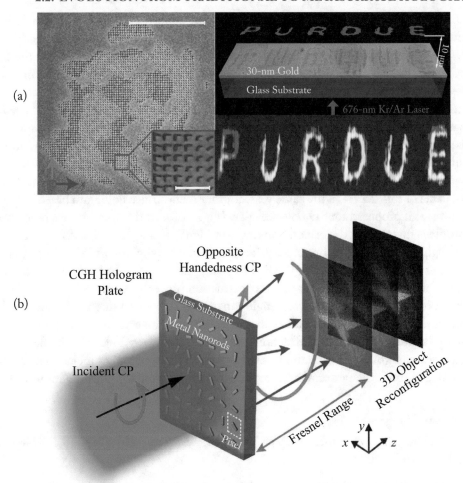

Figure 2.2: The first demonstrations of holography by single-layered metasurfaces. (a) A resonant metasurface hologram implemented by V-shaped meta-atoms [20]. (b) A geometric metasurface hologram implemented by I-shaped meta-atoms [21].

constructed, as the reconstructed infrared (10.6 μm) holographic images shown in the lower panel of Fig. 2.1b.

The phase modulation of metamaterial holography is still based on the accumulated light propagation along the material, the only difference is that metamaterial adopts effective medium constructed by subwavelength resonant structures as a new form of holographic plate rather than the traditional emulsion. Therefore, the metamaterial holograms are still bulky.

Until 2013, metasurface holography was first introduced at an ultra-thin single layer of nano-antennas [20, 21]. The phase modulation in the metasurface no longer depends on long distance propagation of light through a thick media. Instead, it is directly caused by the scattering behavior of the subwavelength meta-atoms, which produce the abrupt phase discontinuity at an ultra-thin interface. Figure 2.2a shows the metasurface hologram implemented at a metallic film carved with V-shaped complementary nano-antennas, the orientation angle and geometry of the nano-antennas were modulated by the amplitude and phase profile of a pre-designed hologram that is calculated from a "PURDUE" pattern. Such a single layered metallic structure with only 30nm thickness works as the ultra-thin hologram for visible light. Within a 676-nm laser illumination of the metasurface, the predesigned holographic image was clearly reconstructed in the right panel of Fig. 2.2a. On the same year, another metasurface hologram based on geometric phase was also proposed for 3D object display (Fig. 2.2b). In this metasurface hologram, the phase profile is mapped by the orientation angle of identical shaped nanorods. For a circularly polarized light illumination of the metasurface, the modulated phase profile will be generated in the output circularly polarized light with opposite handedness. In both kinds of metasurface holograms, the recording and image reconstruction processes were performed without the reference beam, and very large viewing angle ranges were obtained due to the subwavelength pixel size of the hologram. Nevertheless, those two kinds of metasurface holograms were only preliminary demonstrations, the diffraction efficiencies were very low, and the flexibilities to manipulate wavelength and polarization were lacking. In the subsequent metasurface hologram design strategies, both the performances and functionalities were significantly improved.

2.3   REFERENCES

[1] D. Gabor, A new microscopic principle, *Nature*, 161:777–778, 1948. DOI: 10.1038/161777a0. 5

[2] C. Slinger, C. Cameron, and M. Stanley, Computer-generated holography as a generic display technology, *Computer*, 38(8):46–53, 2005. DOI: 10.1109/mc.2005.260. 5

[3] D. P. Kelly, D. S. Monaghan, N. Pandey, T. Kozacki, Micha, A. kiewicz, G. Finke, B. M. Hennelly, and M. Kujawinska, Digital holographic capture and optoelectronic reconstruction for 3D displays, *International Journal of Digital Multimedia Broadcasting*, 2010. DOI: 10.1155/2010/759323. 5

[4] M. Ozaki, J.-i. Kato, and S. Kawata, Surface-plasmon holography with white-light illumination, *Science*, 332(6026):218–220, 2011. DOI: 10.1126/science.1201045. 5, 7

[5] S. Larouche, Y.-J. Tsai, T. Tyler, N. M. Jokerst, and D. R. Smith, Infrared metamaterial phase holograms, *Nat. Mater.*, 11(5):450–454, 2012. DOI: 10.1038/nmat3278. 5, 8

[6] N. Yu and F. Capasso, Flat optics with designer metasurfaces, *Nat. Mater.*, 13(2):139–150, 2014. DOI: 10.1038/nmat3839. 5

[7] Y.-Z. Liu, X.-N. Pang, S. Jiang, and J.-W. Dong, Viewing-angle enlargement in holographic augmented reality using time division and spatial tiling, *Opt. Express*, 21(10):12068–12076, 2013. DOI: 10.1364/oe.21.012068. 6

[8] Y.-Z. Liu, J.-W. Dong, Y.-Y. Pu, H.-X. He, B.-C. Chen, H.-Z. Wang, H. Zheng, and Y. Yu, Fraunhofer computer-generated hologram for diffused: 3D scene in Fresnel region, *Opt. Lett.*, 36(11):2128–2130, 2011. DOI: 10.1364/ol.36.002128. 6, 7

[9] J. W. Goodman, *Introduction to Fourier Optics*, Roberts and Company, 2004. DOI: 10.1063/1.3035549. 6

[10] R. W. Gerchberg and W. O. Saxton, A practical algorithm for the determination of the phase from image and diffraction plane pictures, *Optik, (Jena)*, 35:237, 1972. 7

[11] Y.-Z. Liu, J.-W. Dong, Y.-Y. Pu, B.-C. Chen, H.-X. He, and H.-Z. Wang, High-speed full analytical holographic computations for true-life scenes, *Opt. Express*, 18(4):3345–3351, 2010. DOI: 10.1364/oe.18.003345. 7

[12] J. J. Cowan, The surface plasmon resonance effect in holography, *Opt. Commun.*, 5(2):69–72, 1972. DOI: 10.1016/0030-4018(72)90001-6. 7

[13] S. Maruo, O. Nakamura, and S. Kawata, Evanescent-wave holography by use of surface-plasmon resonance, *Appl. Opt.*, 36(11):2343–2346, 1997. DOI: 10.1364/ao.36.002343. 7

[14] G. P. Wang, T. Sugiura, and S. Kawata, Holography with surface-plasmon-coupled waveguide modes, *Appl. Opt.*, 40(22):3649–3653, 2001. DOI: 10.1364/ao.40.003649. 7

[15] B. Walther, C. Helgert, C. Rockstuhl, F. Setzpfandt, F. Eilenberger, E.-B. Kley, F. Lederer, A. Tünnermann, and T. Pertsch, Spatial and spectral light shaping with metamaterials, *Adv. Mater.*, 24(47):6300–6304, 2012. DOI: 10.1002/adma.201202540. 7, 8

[16] D. R. Smith and N. Kroll, Negative refractive index in left-handed materials, *Phys. Rev. Lett.*, 85(14):2933, 2000. DOI: 10.1103/physrevlett.85.2933. 7

[17] I. Liberal and N. Engheta, Near-zero refractive index photonics, *Nat. Photon.*, 11(3):149–158, 2017. DOI: 10.1038/nphoton.2017.13. 7

[18] M. Choi, S. H. Lee, Y. Kim, S. B. Kang, J. Shin, M. H. Kwak, K.-Y. Kang, Y.-H. Lee, N. Park, and B. Min, A terahertz metamaterial with unnaturally high refractive index, *Nature*, 470(7334):369, 2011. DOI: 10.1038/nature09776. 7

[19] H. Chen, C. T. Chan, and P. Sheng, Transformation optics and metamaterials, *Nat. Mater.*, 9(5):387–396. DOI: 10.1038/nmat2743. 8

[20] X. Ni, A. V. Kildishev, and V. M. Shalaev, Metasurface holograms for visible light, *Nat. Commun.*, 4:3807, 2013. DOI: 10.1038/ncomms3807. 9, 10

[21] L. Huang, X. Chen, H. Mühlenbernd, H. Zhang, S. Chen, B. Bai, Q. Tan, G. Jin, K.-W. Cheah, C.-W. Qiu, J. Li, T. Zentgraf, and S. Zhang, Three-dimensional optical holography using a plasmonic metasurface, *Nat. Commun.*, 4:2808, 2013. DOI: 10.1038/ncomms3808. 9, 10

CHAPTER 3

Phase Modulation Rules of Metasurface Holograms

The versatile and powerful functionalities supported by metasurfaces are fundamentally governed by the underlying phase modulation rules based on the resonance and scattering properties of the individual meta-atoms. Typically, by varying the shape, orientation, size, and position of the meta-atoms, one can apply the resonant phase, geometric P-B phase, propagation phase, detour phase, and any combinations of those phase modulation rules to modulate the wavefront of light versatilely.

3.1 RESONANT PHASE METASURFACES

The building blocks of metamaterials are constructed by subwavelength meta-atoms with proper resonant and scattering properties. There is a π phase jump when the incident wavelength crosses the resonance condition of the meta-atom. To exploit the full 2π phase modulation, the earlier works considering the incident light with linear polarization and extract the output light with crossed linear polarization [1–4]. As shown in Fig. 3.1a, the V-shaped metallic meta-atoms work as subwavelength resonators. By varying the shape of the meta-atom including its opening size and angle, the phase discontinuity originating from the resonance can be achieved at a given wavelength.

The V-shaped resonant metasurfaces can only work at crossed linear polarizations, and the conversion efficiency is very low, although people tried to push the efficiency toward its theoretical limit [5]. The efficiency of metallic resonant metasurfaces can be easily improved to near unitary at reflection mode by the metal-insulator-metal (MIM) structure [6] (Fig. 3.1b). As a single resonator can only produce a phase jump of π, the mirror effect of the MIM structure actually works as double resonators and thus can produce the full 2π phase jump. In addition, the conversion efficiency of those configurations reaches 100% in lower frequencies such as microwave and Terahertz frequencies, where the metallic loss is negligible.

To achieve the simultaneous full 2π phase modulation and high conversion efficiency in the transmission mode, the meta-atom should support both the electric and magnetic resonances with overlapped resonance spectra, which was first demonstrated at metallic structures in microwave frequencies and was referred to as the Huygens metasurfaces [7, 8]. In optical frequencies, Huygens metasurfaces can be implemented in the all-dielectric platform [9]. Figure 3.1c shows the all-dielectric Huygens metasurfaces made of silicon disks with spatially varying sizes.

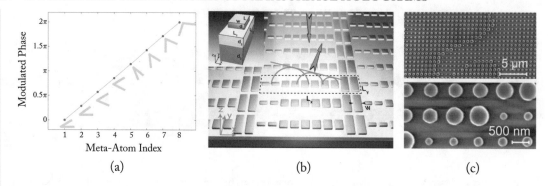

(a)   (b)   (c)

Figure 3.1: Typical metasurface structures based on resonant phase modulation rules. (a) V-shaped metallic antennas with spatially varying shapes that can produce the full 2π phase modulation at the cross-polarization of a linear polarized incident light [5]. (b) Plasmonic nanorods with spatially varying sizes in a reflection mode that can produce the full 2π phase modulation with high-efficiency [6]. (c) Size-varied all-dielectric Mie-scatters that can produce full 2π phase modulations efficiently at the transmission mode [12]. For all those metasurface implementations, the phase modulation is performed near the resonance of the meta-atoms, as a result, the performance of this kind of metasurface is band-limited.

The high-index Silicon disks support multiple Mie scattering modes including any electric multipoles and magnetic multipoles [9–14]. By tailoring the aspect ratio of the all-dielectric disk, electric dipole and magnetic dipole resonances can be readily tuned at will [13]. When the orthogonal electric and magnetic dipole modes overlap with each other with the same resonant strength and damping rate so that the Huygens condition is satisfied, the Silicon disk perform the ultra-high efficiency forward scattering [9, 15, 16]. By arranging such silicon disks as an array with spatially varying diameter [9, 12] or the periodicity [13, 14] of the disk, one can design highly efficient metasurfaces with arbitrary phase profile modulations in the transmission mode.

3.2   GEOMETRIC P-B PHASE METASURFACES

As all types of resonant phase metasurfaces strongly relies on the resonance condition of the meta-atoms, the applicable working frequency range is limited by the resonance bandwidth. Different from resonant phase metasurfaces, the geometric P-B phase metasurfaces are totally decoupled from the spectra response of the meta-atom, and provide an ideal platform for broadband wavefront shaping. When a circular polarized incident light passes through an anisotropic meta-atom, the converted cross component of the output light will be imparted with a geometric P-B phase that is proportional to the orientation angle of the anisotropic meta-atom:

$$\varphi = 2\sigma\psi, \tag{3.1}$$

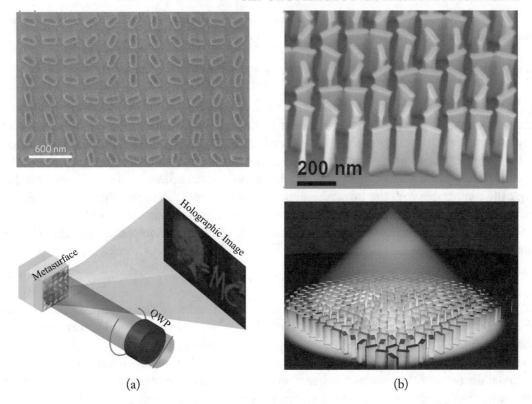

(a)   (b)

Figure 3.2: Geometric P-B phase metasurfaces. (a) High-efficiency reflective geometric meta-surfaces with plasmonic nanorods with spatially varying orientations at a metal-insulator-metal (MIM) configuration [17]. (b) High-efficiency transmissive geometric metasurfaces with all-dielectric high respect ratio nanoposts with spatially varying orientations [19].

where, $\sigma = \pm1$ represents the helicity of the input light, ψ is the orientation angle of the meta-atom, and φ is the geometric P-B phase. The modulated phase has a simple explicit form with respect to the meta-atom parameter, largely simplifying the design procedure for a given func-tionality. The only needed optimization process of geometric phase metasurfaces should be the parameter optimization for high conversion efficiency between the crossed circular polarizations by full-wave simulations of a practical plasmonic or all-dielectric structure. When the geometric parameters of the anisotropic meta-atom are optimized to behave like a half wave-plate, the con-version efficiency of geometric metasurfaces can theoretically reach 100% for a broad bandwidth. In practical experiment, the high-efficiency geometric phase metasurfaces can be implemented either in the MIM plasmonic platform for reflective wavefront shaping (Fig. 3.2a), or in the all-dielectric platform for transmissive wavefront shaping (Fig. 3.2b). Ultra-high diffraction efficiency of 80% for meta-hologram [17] and diffraction-limited high-efficiency meta-lens are successfully demonstrated [18, 19].

3.3 PROPAGATION PHASE METASURFACES

In the all-dielectric platform of metasurface, the high respect ratio nanopost not only support the geometric P-B phase induced by the orientation angle of the anisotropic structure, but also support the propagation phase that is originated from the waveguide mode propagation inside the high-index nanopost. As the refractive index of nanopost is much higher than the background, the required waveguide length (which determines the thickness of the metasurfaces) is typically comparable with the working wavelength in freespace. The spatially varying profile of propagation phase is realized by tuning the cross section size of the nanopost, which changes the effective index of the waveguide mode, and thus tunes the propagation phase. The expression of propagation phase can be written as

$$\varphi = \frac{\omega}{c} n_{eff} h, \tag{3.2}$$

where h is the height of the nanopost, n_{eff} is the effect index of the waveguide mode, ω is the angular frequency of light, c is the light speed constant, and φ is the propagation phase. As shown in Fig. 3.3, cylinder waveguides are employed to modulate the propagation phase without the polarization dependence. At the given working wavelength, the propagation phase increases with the diameter of the circular cross-section, as the effective index of the waveguide mode at that wavelength increases with the waveguide size. At some specific diameters, there are sharp peaks associated with the transmitted amplitude due to the Fabry–Perot resonance, and those parameters should be avoided in the real metasurface design.

3.4 DETOUR PHASE METASURFACES

The detour phase is an old phase modulation approach in traditional holography construction [22, 23]. However, traditional detour phase holography is based on the interference of light through small apertures, which has many unwanted side effect such as twin image, strong zero-order-diffraction, and low diffraction efficiency. Recently, detour phase was revisited in the metasurface community with significantly improved performances [24–28]. Meta-atoms with desired scattering properties can be employed to replace traditional light aperture. Arranging such meta-atoms in a metagrating configuration [29–32] can lead to near-unitary diffraction efficiency in the high diffraction orders [33–38]. According to the detour phase modulation rule, meta-atoms can be considered as effective apertures, whose relative displacements can be directly mapped to the detour phase with the following explicit expression:

$$\varphi = \frac{2\pi p}{p_0}, \tag{3.3}$$

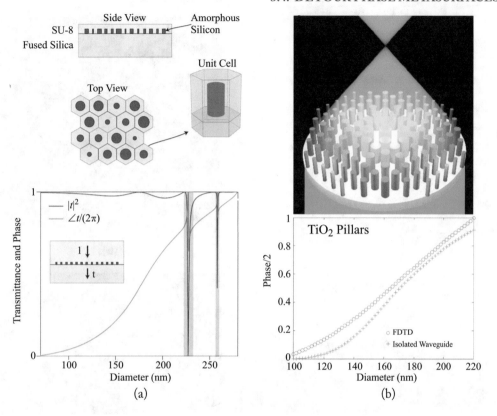

Figure 3.3: Propagation phase based metasurfaces that modulate the phase profile of a wavefront by the effective index of the vertical dielectric waveguides with spatially varying cross sections, which are typically realized by high refractive index dielectric materials such as (a) Silicon [20] and (b) Titanium Dioxide [21].

where p is the relative displacement of the meta-atom, p_0 is the periodicity of the metasurface, and φ is the modulated detour phase by the meta-atom. As shown in Fig. 3.4a,b, the silicon tri-bar meta-atoms are considered as the apertures of detour phase. Periodic tri-bar metagrating can totally channel the incident light into ±1 diffraction orders, with complete suppression of the zero-diffraction order. Spatially varying relative displacements of such tri-bar meta-atoms according to the detour phase modulation rule, broadband dispersionless holographic images can be efficiently produced in the ±1 diffraction orders [25].

   Actually, the detour phase modulation rule can be generalized to arbitrary incident angles and full space including not only transmission space but also reflection space with high diffraction efficiencies. In the MIM configuration as shown in Fig. 3.4c,d, plasmonic nanorods metagratings perform near-unitary diffraction efficiency at a single diffraction order within the

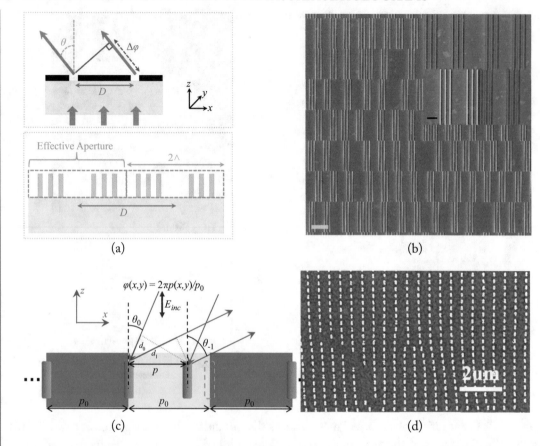

Figure 3.4: Detour phase based metagratings that can shape the wavefront in a broadband and wide-angle range. (a, b) All-dielectric metagratings with tri-bar unit cells modulated by the detour phase [25]. (c, d) Plasmonic nanorods with spatially varying positions forms a high-efficient plasmonic metagrating based on detour phase [39].

extraordinary optical diffraction (EOD) regime [39]. The detour phase is also applicable for such metagratings under oblique incidence (Fig. 3.4c), and high-efficient broadband holographic images can be generated with extremely large angle tolerance in such plasmonic metagrating platform [39].

3.5   METASURFACES COMBINING GEOMETRIC P-B PHASE AND PROPAGATION PHASE

To integrate more functionalities at a single metasurface layer, the combination of the above four basic types of phase modulation rules should be exploited. The combination of geometric P-B

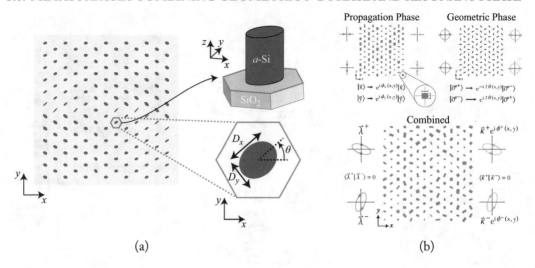

(a)   (b)

Figure 3.5: Metasurfaces with combined geometric PB phase and propagation phase modulation by simultaneously tailoring the size and orientation of the all-dielectric (a) elliptical [40] and (b) rectangular nanopost array [41, 42], which can perform the high-efficiency full control of phase and polarization states of arbitrary wavefront.

phase and propagation phase in the all-dielectric metasurface is a powerful strategy to extend its functionalities. This type of metasurfaces simultaneously tailors the size parameter in two orthogonal directions as well as the orientation angle of an anisotropic all-dielectric nanoposts. The cross sections of such anisotropic nanoposts could be elliptical (Fig. 3.5a) [40] or rectangle (Fig. 3.5b) [41, 42]. The propagation phases in two orthogonal direction are mapped to the size parameters at corresponding directions, while the geometric P-B phase is mapped directly to the orientation angle. Jones matrix analysis indicates that both phase and polarization states of incident light can be simultaneously and independently modulated by this strategy, which promises versatile functionalities including: arbitrary polarization multiplexing [41], arbitrary spin to orbital angular momentum conversion [42], and so on.

3.6 METASURFACES COMBINING GEOMETRIC P-B PHASE AND RESONANT PHASE

The geometric P-B phase can also be combined with the resonant phase to engineer the phase dispersion of broadband light. The main issue of metalens is the chromatic behavior, i.e., the incident light with different wavelengths will be focused at different positions, which dramatically degrades the imaging quality of colorful object in the visible frequencies. Introducing the so-called integrated-resonant unit element (IRUE) as the meta-atom, resonant phase dispersion over a broadband in near infrared [43] and visible [44] frequency range could be engineered at

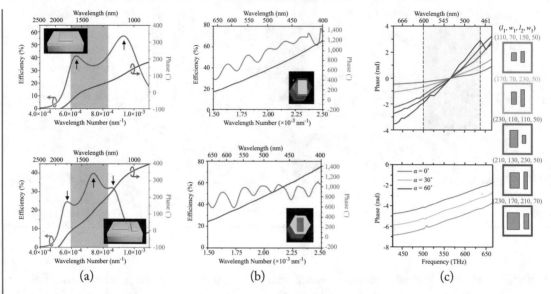

Figure 3.6: Metasurface with combined geometric phase and resonance phase to simultaneously control the phase and phase dispersion of light. (a) Metasurface tailoring the orientation angle of plasmonic integrated resonators in a MIM configuration, leading to achromatic metalens in the reflection mode [43]. (b) Metasurfaces tailoring the orientation angle of multiple all-dielectric resonators with solid/hollow structure [44], and double bar structure [45], leading to achromatic metalens in the transmission mode.

will by tailoring the inclusions of IRUE, which is composed of multiple plasmonic (Fig. 3.6a) or all-dielectric (Fig. 3.6b,c) nanoscatters in each unit cell. The other degree of freedom, namely, orientation angle of the IRUE can be independently tailored to generate the required geometric phase profile for focusing at the designed central wavelength. If only the geometric phase is employed to construct the required phase profile for focusing, the focusing will be chromatic. Together with the phase dispersion engineering over the required bandwidth, phase compensation by the resonance of IRUE will lead to the achromatic focusing, which channel the incident light of different wavelengths into the same focusing point.

3.7   METASURFACES COMBINING GEOMETRIC P-B PHASE AND RESONANT SPECTRA OF META-ATOMS

This kind of metasurfaces records the wavefront information by geometric P-B phase with spatially varying orientations of the meta-atoms. At the same time, the size parameter of meta-atoms that corresponds to resonant spectrum is tailored to control the frequency or amplitude information of the wavefront. On one hand, the diffraction resonant peaks for different-sized

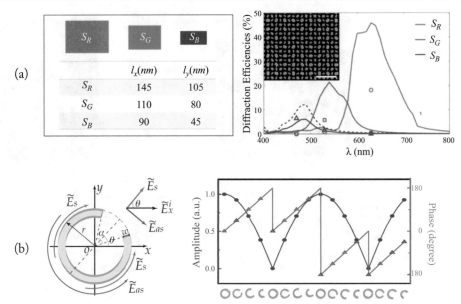

(a)

(b)

Figure 3.7: Metasurfaces combing geometric P-B phase and resonant spectrum response of the meta-atom. (a) Full-color meta-holograms by combing geometric P-B phase and resonant spectrum of meta-atom [46]. The size parameter of the meta-atom determines the spectrum response, while the orientation parameter of the meta-atom modulate the phase. (b) Complete phase and amplitude modulations by combining geometric P-B phase and spectra response of a split-ring resonator (SPR) [4]. The orientation angle and opening size of the SPR are simultaneously controlled to modulate both the amplitude and phase.

meta-atoms locate at different wavelengths. Interleaving multiple-sized meta-atoms in each unit cell, each set of meta-atoms carries independent holographic information for different wavelengths. Full-color holographic imaging could be constructed in this way (Fig. 3.7a) [46]. On the other hand, different resonant spectra will lead to different amplitudes at a given wavelength, and thus the combing of geometric phase modulation and resonant spectra of the meta-atoms can simultaneously and independently control the amplitude and phase of the wavefront (Fig. 3.7b) [4].

3.8   METASURFACES COMBINING GEOMETRIC P-B PHASE AND DETOUR PHASE

In a diatomic arrangement of identical meta-atoms (Fig. 3.8), multiple degrees of freedom including orientation angle ψ_1, ψ_2 and relative displacement p_1, p_2 of both meta-atoms can be full exploited to manage the geometric P-B phase and detour phase simultaneously. As both the

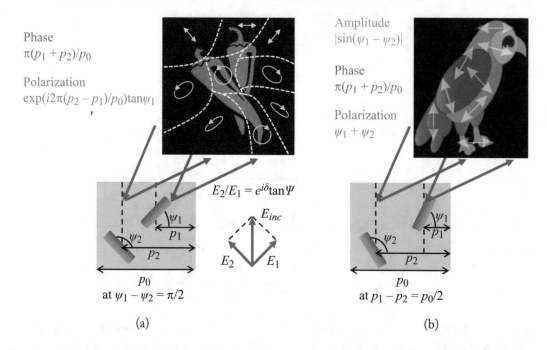

Phase
$\pi(p_1 + p_2)/p_0$

Polarization
$\exp(i2\pi(p_2 - p_1)/p_0)\tan\psi_1$

Amplitude
$|\sin(\psi_1 - \psi_2)|$

Phase
$\pi(p_1 + p_2)/p_0$

Polarization
$\psi_1 + \psi_2$

$E_2/E_1 = e^{i\delta}\tan\Psi$

(a)   (b)

Figure 3.8: Metasurfaces combing both geometric P-B phase and detour phase. Both the orientation angle (ψ_1, ψ_2) and relative displacement (p_1, p_2) of the diatomic meta-molecule are tuned to simultaneously modulate multiple attributes of light wavefront. (a) At the condition $\psi_1 - \psi_2 = \pi/2$, the metasurface can simultaneously modulate the phase, polarization state (including any elliptical polarizations) with explicit expressions $\pi(p_1 + p_2)/p_0$, $(\cos\psi, e^{i2\pi(p_2-p_1)/p_0}\sin\psi_1)^T$, respectively [47]. (b) At the condition $p_1 - p_2 = p_0/2$, the amplitude, phase, and polarization orientation of light wave can be simultaneously and independently modulated through explicit expressions: $|\sin(\psi_1 - \psi_2)|$, $\pi(p_1 + p_2)/p_0$, and $\begin{pmatrix} \cos(\psi_1 + \psi_2) \\ \sin(\psi_1 + \psi_2) \end{pmatrix}$, respectively [48]. For both cases, the modulation relationships between the attributes of light field and the parameters of the structure are irrespective of both incident angle and wavelength, because the geometric P-B phase and detour phase are totally decoupled from the spectral response.

geometric P-B phase and detour phase are irrespective of wavelengths, metasurface based on this strategy can control multiple attributes of light in a broadband. There are two special cases in the diatomic metasurface design that have explicit Jones matrix formulas.

1. When the two meta-atoms are orthogonally arranged with each other ($\psi_1 - \psi_2 = \pi/2$) (Fig. 3.8a), the Jones matrix of the diatomic meta-molecule for the diffraction light could

be written as

$$\mathbf{J}(p_1, p_2, \psi_1, \psi_2) = C\mathbf{R}^{-1}(\psi_1) \begin{pmatrix} e^{i2\pi p_1/p_0} & 0 \\ 0 & e^{i2\pi p_2/p_0} \end{pmatrix} \mathbf{R}(\psi_1),$$

where C is the coupling constant between the diffracted and incident light, $\mathbf{R}$ is the rotational matrix. In this case, assuming a linearly polarized input light with Jones vector $E_{in} = (0, 1)^T$, both the phase and polarization state of light can be completely modulated through explicit forms $\pi(p_1 + p_2)/p_0$, $(\cos \psi, e^{i2\pi(p_2-p_1)/p_0} \sin \psi_1)^T$, respectively, which are independent of both wavelengths and incident angles.

2. When the relative displacement of the two meta-atoms equals to the half periodicity of the metasurface ($p_1 - p_2 = p_0/2$) (Fig. 3.8b), the Jones matrix can be written as

$$\mathbf{J}(p_1, p_2, \psi_1, \psi_2)$$
$$= iC \sin(\psi_1 - \psi_2) e^{i\pi(p_1+p_2)/p_0} \begin{pmatrix} -\sin(\psi_1 + \psi_2) & \cos(\psi_1 + \psi_2) \\ \cos(\psi_1 + \psi_2) & \sin(\psi_1 + \psi_2) \end{pmatrix},$$

where C is the coupling constant between diffracted and incident light. Assuming a linearly polarized input light with Jones vector $E_{in} = (0, 1)^T$, the amplitude, phase, and polarization orientation of output light can be all controlled through explicit expressions at the same time through explicit forms: $|\sin(\psi_1 - \psi_2)|$, $\pi(p_1 + p_2)/p_0$, and $\begin{pmatrix} \cos(\psi_1 + \psi_2) \\ \sin(\psi_1 + \psi_2) \end{pmatrix}$, respectively. In addition, the modulation of the three attributes of light can be performed for a broadband and wide-angle range, because the modulation expressions are all independent of both incident angle and wavelengths.

3.9   REFERENCES

[1] N. Yu, P. Genevet, M. A. Kats, F. Aieta, J.-P. Tetienne, F. Capasso, and Z. Gaburro, Light propagation with phase discontinuities: Generalized laws of reflection and refraction, *Science*, 334(6054):333–337, 2011. DOI: 10.1126/science.1210713. 13

[2] F. Zhou, Y. Liu, and W. Cai, Plasmonic holographic imaging with V-shaped nanoantenna array, *Opt. Express*, 21(4):4348–4354, 2013. DOI: 10.1364/oe.21.004348.

[3] X. Zhang, Z. Tian, W. Yue, J. Gu, S. Zhang, J. Han, and W. Zhang, Broadband terahertz wave deflection based on C-shape complex metamaterials with phase discontinuities, *Adv. Mater.*, 25(33):4567–4572, 2013. DOI: 10.1002/adma.201204850.

[4] L. Liu, X. Zhang, M. Kenney, X. Su, N. Xu, C. Ouyang, Y. Shi, J. Han, W. Zhang, and S. Zhang, Broadband metasurfaces with simultaneous control of phase and amplitude, *Adv. Mater.*, 26(29):5031–5036, 2014. DOI: 10.1002/adma.201401484. 13, 21

[5] F. Qin, L. Ding, L. Zhang, F. Monticone, C. C. Chum, J. Deng, S. Mei, Y. Li, J. Teng, M. Hong, S. Zhang, A. Alù, and C.-W. Qiu, Hybrid bilayer plasmonic metasurface efficiently manipulates visible light, *Sci. Adv.*, 2(1):e1501168, 2016. DOI: 10.1126/sciadv.1501168. 13, 14

[6] S. Sun, K.-Y. Yang, C.-M. Wang, T.-K. Juan, W. T. Chen, C. Y. Liao, Q. He, S. Xiao, W.-T. Kung, G.-Y. Guo, L. Zhou, and D. P. Tsai, High-efficiency broadband anomalous reflection by gradient meta-surfaces, *Nano Lett.*, 12(12):6223–6229, 2012. DOI: 10.1021/nl3032668. 13, 14

[7] C. Pfeiffer and A. Grbic, Metamaterial huygens' surfaces: Tailoring wave fronts with reflectionless sheets, *Phys. Rev. Lett.*, 110(19):197401, 2013. DOI: 10.1103/physrevlett.110.197401. 13

[8] K. Chen, Y. Feng, F. Monticone, J. Zhao, B. Zhu, T. Jiang, L. Zhang, Y. Kim, X. Ding, S. Zhang, A. Alù, and C.-W. Qiu, A reconfigurable active huygens' metalens, *Adv. Mater.*, 29(17):1606422–n/a, 2017. DOI: 10.1002/adma.201606422. 13

[9] M. Decker, I. Staude, M. Falkner, J. Dominguez, D. N. Neshev, I. Brener, T. Pertsch, and Y. S. Kivshar, High-efficiency dielectric huygens' surfaces, *Adv. Opt. Mater.*, 3(6):813–820, 2015. DOI: 10.1002/adom.201400584. 13, 14

[10] I. Staude, A. E. Miroshnichenko, M. Decker, N. T. Fofang, S. Liu, E. Gonzales, J. Dominguez, T. S. Luk, D. N. Neshev, I. Brener, and Y. Kivshar, Tailoring directional scattering through magnetic and electric resonances in subwavelength silicon nanodisks, *ACS Nano*, 7(9):7824–7832, 2013. DOI: 10.1021/nn402736f.

[11] J. Sautter, I. Staude, M. Decker, E. Rusak, D. N. Neshev, I. Brener, and Y. S. Kivshar, Active tuning of all-dielectric metasurfaces, *ACS Nano*, 9(4):4308–4315, 2015. DOI: 10.1021/acsnano.5b00723.

[12] L. Wang, S. Kruk, H. Tang, T. Li, I. Kravchenko, D. N. Neshev, and Y. S. Kivshar, Grayscale transparent metasurface holograms, *Optica*, 3(12):1504–1505, 2016. DOI: 10.1364/optica.3.001504. 14

[13] K. E. Chong, I. Staude, A. James, J. Dominguez, S. Liu, S. Campione, G. S. Subramania, T. S. Luk, M. Decker, D. N. Neshev, I. Brener, and Y. S. Kivshar, Polarization-independent silicon metadevices for efficient optical wavefront control, *Nano Lett.*, 15(8):5369–5374, 2015. DOI: 10.1021/acs.nanolett.5b01752. 14

[14] K. E. Chong, L. Wang, I. Staude, A. R. James, J. Dominguez, S. Liu, G. S. Subramania, M. Decker, D. N. Neshev, I. Brener, and Y. S. Kivshar, Efficient polarization-insensitive complex wavefront control using huygens' metasurfaces based on dielectric resonant metaatoms, *ACS Photon.*, 3(4):514–519, 2016. DOI: 10.1021/acsphotonics.5b00678. 14

[15] W. Zhao, H. Jiang, B. Liu, J. Song, and Y. Jiang, High-efficiency beam manipulation combining geometric phase with anisotropic huygens surface, *Appl. Phys., Lett.*, 108(18):181102, 2016. DOI: 10.1063/1.4948518. 14

[16] W. Zhao, H. Jiang, B. Liu, J. Song, Y. Jiang, C. Tang, and J. Li, Dielectric huygens' metasurface for high-efficiency hologram operating in transmission mode, *Sci. Rep.*, 6:30613, 2016. DOI: 10.1038/srep30613. 14

[17] G. Zheng, H. Mühlenbernd, M. Kenney, G. Li, T. Zentgraf, and S. Zhang, Metasurface holograms reaching: 80% efficiency, *Nat. Nanotechnol.*, 10(4):308–312, 2015. DOI: 10.1038/nnano.2015.2. 15, 16

[18] M. Khorasaninejad, W. T. Chen, R. C. Devlin, J. Oh, A. Y. Zhu, and F. Capasso, Metalenses at visible wavelengths: Diffraction-limited focusing and subwavelength resolution imaging, *Science*, 352(6290):1190–1194, 2016. DOI: 10.1126/science.aaf6644. 16

[19] H. Liang, Q. Lin, X. Xie, Q. Sun, Y. Wang, L. Zhou, L. Liu, X. Yu, J. Zhou, T. F. Krauss, and J. Li, Ultrahigh numerical aperture metalens at visible wavelengths, *Nano Lett.*, 18(7):4460–4466, 2018. DOI: 10.1021/acs.nanolett.8b01570. 15, 16

[20] A. Arbabi, E. Arbabi, Y. Horie, S. M. Kamali, and A. Faraon, Planar metasurface retroreflector, *Nat. Photon.*, 11(7):415–420, 2017. DOI: 10.1038/nphoton.2017.96. 17

[21] M. Khorasaninejad, A. Y. Zhu, C. Roques-Carmes, W. T. Chen, J. Oh, I. Mishra, R. C. Devlin, and F. Capasso, Polarization-insensitive metalenses at visible wavelengths, *Nano Lett.*, 16(11):7229–7234, 2016. DOI: 10.1021/acs.nanolett.6b03626. 17

[22] B. R. Brown and A. W. Lohmann, Complex spatial filtering with binary masks, *Appl. Opt.*, 5(6):967–969, 1966. DOI: 10.1364/ao.5.000967. 16

[23] W.-H. Lee, Binary computer-generated holograms, *Appl. Opt.*, 18(21):3661–3669, 1979. DOI: 10.1364/ao.18.003661. 16

[24] C. Min, J. Liu, T. Lei, G. Si, Z. Xie, J. Lin, L. Du, and X. Yuan, Plasmonic nano-slits assisted polarization selective detour phase meta-hologram, *Laser Photon. Rev.*, 10(6):978–985, 2016. DOI: 10.1002/lpor.201600101. 16

[25] M. Khorasaninejad, A. Ambrosio, P. Kanhaiya, and F. Capasso, Broadband and chiral binary dielectric meta-holograms, *Sci. Adv.*, 2(5):e1501258, 2016. DOI: 10.1126/sciadv.1501258. 17, 18

[26] J. Lin, P. Genevet, M. A. Kats, N. Antoniou, and F. Capasso, Nanostructured holograms for broadband manipulation of vector beams, *Nano Lett.*, 13(9):4269–4274, 2013. DOI: 10.1021/nl402039y.

[27] Z. Xie, T. Lei, G. Si, X. Wang, J. Lin, C. Min, and X. Yuan, Meta-holograms with full parameter control of wavefront over a: 1000 nm bandwidth, *ACS Photon.*, 4(9):2158–2164, 2017. DOI: 10.1021/acsphotonics.7b00710.

[28] Z.-L. Deng, X. Ye, Y.-Y. Qiu, Q.-A. Tu, T. Shi, Z.-P. Zhuang, Y. Cao, B.-O. Guan, N. Feng, G. P. Wang, A. Alù, J.-W. Dong, and X. Li, Transmissive metragrating for arbitray wavefront shaping over the full visible spectrum, *arXiv:2003.08036*, 2020. 16

[29] Z.-L. Deng, S. Zhang, and G. P. Wang, A facile grating approach towards broadband, wide-angle and high-efficiency holographic metasurfaces, *Nanoscale*, 8(3):1588–1594, 2016. DOI: 10.1039/c5nr07181j. 16

[30] Y. Ra'di, D. L. Sounas, and A. Alù, Metagratings: Beyond the limits of graded metasurfaces for wave front control, *Phys. Rev. Lett.*, 119(6):067404, 2017. DOI: 10.1103/physrevlett.119.067404.

[31] Z.-L. Deng, S. Zhang, and G. P. Wang, Wide-angled off-axis achromatic metasurfaces for visible light, *Opt. Express*, 24(20):23118–23128, 2016. DOI: 10.1364/oe.24.023118.

[32] Z.-L. Deng, Y. Cao, X. Li, and G. P. Wang, Multifunctional metasurface: From extraordinary optical transmission to extraordinary optical diffraction in a single structure, *Photon. Res.*, 6(5):443–450, 2018. DOI: 10.1364/prj.6.000659. 16

[33] Y. Ra'di and A. Alù, Reconfigurable metagratings, *ACS Photon.*, 5(5):1779–1785, 2018. DOI: 10.1021/acsphotonics.7b01528. 16

[34] V. Popov, F. Boust, and S. N. Burokur, Controlling diffraction patterns with metagratings, *Phys. Rev. Appl.*, 10(1):011002, 2018. DOI: 10.1103/physrevapplied.10.011002.

[35] Z. Fan, M. R. Shcherbakov, M. Allen, J. Allen, B. Wenner, and G. Shvets, Perfect diffraction with multiresonant bianisotropic metagratings, *ACS Photon.*, 5(11):4303–4311, 2018. DOI: 10.1021/acsphotonics.8b00434.

[36] D. Sell, J. Yang, S. Doshay, R. Yang, and J. A. Fan, Large-angle, multifunctional metagratings based on freeform multimode geometries, *Nano Lett.*, 17(6):3752–3757, 2017. DOI: 10.1021/acs.nanolett.7b01082.

[37] A. Epstein and O. Rabinovich, Unveiling the properties of metagratings via a detailed analytical model for synthesis and analysis, *Phys. Rev. Appl.*, 8(5):054037, 2017. DOI: 10.1103/physrevapplied.8.054037.

[38] M. Khorasaninejad and F. Capasso, Broadband multifunctional efficient meta-gratings based on dielectric waveguide phase shifters, *Nano Lett.*, 15(10):6709–6715, 2015. DOI: 10.1021/acs.nanolett.5b02524. 16

[39] Z.-L. Deng, J. Deng, X. Zhuang, S. Wang, T. Shi, G. P. Wang, Y. Wang, J. Xu, Y. Cao, X. Wang, X. Cheng, G. Li, and X. Li, Facile metagrating holograms with broadband and extreme angle tolerance, *Light Sci. Appl.*, 7(1):78, 2018. DOI: 10.1038/s41377-018-0075-0. 18

[40] A. Arbabi, Y. Horie, M. Bagheri, and A. Faraon, Dielectric metasurfaces for complete control of phase and polarization with subwavelength spatial resolution and high transmission, *Nat. Nanotechnol.*, 10(11):937–943, 2015. DOI: 10.1038/nnano.2015.186. 19

[41] J. P. Balthasar Mueller, N. A. Rubin, R. C. Devlin, B. Groever, and F. Capasso, Metasurface polarization optics: Independent phase control of arbitrary orthogonal states of polarization, *Phys. Rev. Lett.*, 118(11):113901, 2017. DOI: 10.1103/physrevlett.118.113901. 19

[42] R. C. Devlin, A. Ambrosio, N. A. Rubin, J. P. B. Mueller, and F. Capasso, Arbitrary spin-to-orbital angular momentum conversion of light, *Science*, 358:896–901, 2017. DOI: 10.1126/science.aao5392. 19

[43] S. Wang, P. C. Wu, V.-C. Su, Y.-C. Lai, C. Hung Chu, J.-W. Chen, S.-H. Lu, J. Chen, B. Xu, C.-H. Kuan, T. Li, S. Zhu, and D. P. Tsai, Broadband achromatic optical metasurface devices, *Nat. Commun.*, 8(1):187, 2017. DOI: 10.1038/s41467-017-00166-7. 19, 20

[44] S. Wang, P. C. Wu, V.-C. Su, Y.-C. Lai, M.-K. Chen, H. Y. Kuo, B. H. Chen, Y. H. Chen, T.-T. Huang, J.-H. Wang, R.-M. Lin, C.-H. Kuan, T. Li, Z. Wang, S. Zhu, and D. P. Tsai, A broadband achromatic metalens in the visible, *Nat. Nanotechnol.*, 13(3):227–232, 2018. DOI: 10.1038/s41565-017-0052-4. 19, 20

[45] W. T. Chen, A. Y. Zhu, V. Sanjeev, M. Khorasaninejad, Z. Shi, E. Lee, and F. Capasso, A broadband achromatic metalens for focusing and imaging in the visible, *Nat. Nanotechnol.*, 13(3):220–226, 2018. DOI: 10.1038/s41565-017-0034-6. 20

[46] B. Wang, F. Dong, Q.-T. Li, D. Yang, C. Sun, J. Chen, Z. Song, L. Xu, W. Chu, Y.-F. Xiao, Q. Gong, and Y. Li, Visible-frequency dielectric metasurfaces for multiwavelength achromatic and highly dispersive holograms, *Nano Lett.*, 16(8):5235–5240, 2016. DOI: 10.1021/acs.nanolett.6b02326. 21

[47] Z.-L. Deng, J. Deng, X. Zhuang, S. Wang, K. Li, Y. Wang, Y. Chi, X. Ye, J. Xu, G. P. Wang, R. Zhao, X. Wang, Y. Cao, X. Cheng, G. Li, and X. Li, Diatomic metasurface for vectorial holography, *Nano Lett.*, 18(5):2885–2892, 2018. DOI: 10.1021/acs.nanolett.8b00047. 22

[48] Z.-L. Deng, M. Jin, X. Ye, S. Wang, T. Shi, J. Deng, N. Mao, Y. Cao, B.-O. Guan, A. Alù, G. Li, and X. Li, Full-color complex-amplitude vectorial holograms based on multi-freedom metasurfaces, *Adv. Fun. Mater.*, 2020. DOI: 10.1002/adfm.201910610. 22

CHAPTER 4

Metasurface Polarization Holography

At the time when holography was invented, the "holo" information of a light wavefront represent both the intensity and phase information. Actually, the polarization states of light are also basic attribute of light. In earlier times, holograms are recorded in refractive bulky materials that are typically polarization insensitive, so polarization states of light wavefront were not effectively manipulated and most of the optical diffraction phenomena were treated by scalar diffraction theory [1]. To fully describe the light wavefront as a vector field, polarization degree of freedom must be further manipulated. In traditional optics, polarization holography was realized by employing natural photo-induced anisotropic materials, such as the Azobenzene materials [2]. In this way, polarization multiplexed dual holographic images were demonstrated in the traditional way [3–5]. However, such polarization holography can only be recorded by the optical interference setup, capable of only limited manipulation of polarization states of light.

In the metasurface platform, the highly controllable polarization response of meta-atoms provide a versatile way to manipulate the polarization state of a light wavefront [6–10]. Metasurfaces are capable for linear polarization multiplexing holograms that can reconstruct different images for crossed linear polarizations, helicity multiplexed holograms that can reconstruct different holographic images for different helicity of the circular polarized light, and even the vectorial holography that can exhibit arbitrary polarization distributions over a holographic image.

4.1 LINEAR POLARIZATION MULTIPLEXED METASURFACE HOLOGRAPHY

Anisotropic meta-atoms with orthogonal orientations are the simplest way to realize polarization multiplexing. The plasmonic resonance of metallic nanorods are highly dependent on the polarization of the excitation light. Typically, only the light with polarization parallel to the long axis of the nanorod can excite the localized surface plasmon resonance. By placing two sets of perpendicular nanorods, each set of nanorods will independently response to two orthogonal linear polarizations of incident light. At the same time, resonance phases of each sets of nanorods are employed to record two independent holographic images. As both the modulated resonance phase and polarization response are dependent on the size and shape parameters of

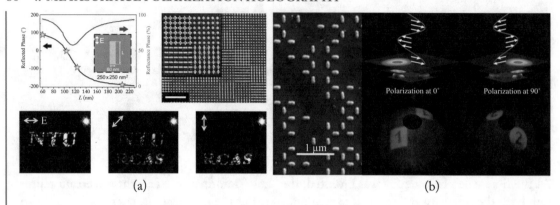

Figure 4.1: Linear polarization multiplexing in resonant phase metasurfaces. Both the polarization response and phase modulation are dependent on the resonant properties of the meta-atom. As a result, the phase levels for each polarization states are largely limited. (a) A four-phase-level polarization multiplexing metasurface made of Aluminum bar-shaped and cross-shaped meta-atoms [11]. (b) A two-phase-level polarization multiplexing metasurface made of silver nanorods with perpendicular orientations [12].

the meta-atoms, the modulated phase levels are highly limited. In Fig. 4.1a, four phase levels near the plasmonic resonance are employed to record the holographic images, and multiple sets of vertical, horizontal, and crossed nanorods are mixed together to realize the linear polarization multiplexing [11]. By simply recording the holographic images through a binary modulation scheme (Fig. 4.1b), one can also realize the linear polarization multiplexed holography. However, due to the limited phase level, twin images and strong zero-order diffraction were always present in the reconstructed holographic images [12].

In the all-dielectric metasurface, polarization multiplexing can be realized with high performances. The meta-atoms of propagation phase all-dielectric metasurfaces are composed of high respect-ratio nanoposts. The nanoposts with elliptical or rectangular cross-sections have anisotropic responses to incident light. In other words, the waveguide propagation phase produced by those nanoposts can be independently tailored by the dimension parameters along the long axis and short axis, respectively. In this way, polarization-dependent deflection (Fig. 4.2a), and polarization multiplexed holographic images (Fig. 4.2b) can be highly efficiently performed [13].

In addition to the linear polarization multiplexing implemented in resonant phase and propagation phase metasurfaces, detour phase meta-holograms provide a robust platform for polarization multiplexing, as the detour phase is completely decoupled from the polarization and spectrum response of the meta-atoms. In Fig. 4.3a, two sets of detour phases are recorded by the horizontal and vertical metallic slit arrays, which response to two perpendicular linearly polarized incident light. In this way, the predesigned vortex beam and Airy beam can be gener-

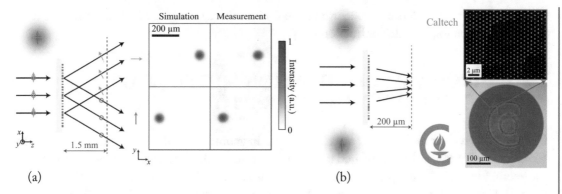

(a)   (b)

Figure 4.2: Linear polarization multiplexing in propagation metasurfaces. In the propagation metasurfaces made of anisotropic all-dielectric nanoposts, the propagation phases along the fast-axis and slow-axis of the anisotropic nanoposts can be independently tailored by the width and length of the rectangular or elliptical cross-section of the nanoposts [13].

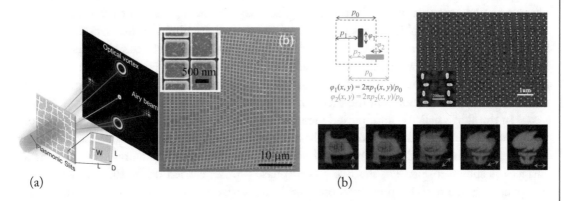

(a)   (b)

Figure 4.3: Linear polarization multiplexing in detour phase metasurfaces. The detour phase is determined by the position of the meta-atom, which is totally decoupled from the polarization and spectral response of the meta-atoms. As a result, continuous phase levels could be realized in the detour phase multiplexing, as demonstrated in the (a) metallic slit structure [14] and the metallic nanorod structure [16].

ated by the horizontally and vertically polarized light illuminations [14]. The traditional detour phase holograms were always accompanied by twin images and strong zero-order diffraction. In the EOD metagrating configuration [15], twin images and strong zero-order diffraction can be effectively suppressed by localized surface plasmon mode. In an EOD metagrating with orthogonal plasmonic nanorods as shown in Fig. 4.3b, the relative positions of the two sets of nanorods can be independently tailored to recording the *boat* and *torch* images, which were free of twin images and strong zero-order diffraction [16].

4.2   CIRCULAR POLARIZATION MULTIPLEXED METASURFACE HOLOGRAPHY

Circular polarization multiplexing are very popular in geometric P-B metasurfaces, because the geometric P-B phase is directly related to the circular polarization of light. As shown in Eq. (3.1), the geometric P-B phase is proportional to the orientation angle and the helicity of the incident light. For incident lights with opposite helicity, the modulated geometric phase will flip its sign. It means that, if we design a holographic image for LCP, the conjugated holographic image will appear upon the illumination of RCP, or vice versa. Figure 4.4a shows the recording of a flower and a bee image that are separately placed on two sides of the incident light. With LCP illumination, the flower was reconstructed on the left side, and the bee was reconstructed on the right side. With RCP illumination, the places of the flower and bee exchanged [17]. If we observe the holographic images in a specific place, say, in the right side of the incident light beam, different holographic images appear for different helicity illuminations. Similar helicity-multiplexed holograms are also implemented by a dielectric metasurfaces [18]. Based on the helicity multiplexing, one can also design helicity dependent multiple functionalities as shown in Fig. 4.4b. For LCP illumination, the metasurface generate a "cat"-shaped holographic image, while for the RCP illumination, the metasurface works as a focusing lens [19].

In the above helicity-multiplexed metasurfaces, the phase profile recorded by the two orthogonal circular polarizations are actually not independent. The phase profile carried on by LCP is always the opposite of that carried on by RCP. In the following chiral geometric metasurfaces as shown in Fig. 4.5, the two sets of phase profiles carried on by RCP and LCP are independently from each other. In those chiral metasurfaces, the asymmetric meta-atom has chiral responses in the reflection (Fig. 4.5a–c) [20] or transmission (Fig. 4.5d–f) [21] spectra. In Fig. 4.5a, the double-split ring resonators (DSPR) in a MIM configuration can reflect the incident light with negative helicity, while the light with the opposite helicity is completely absorbed at the resonance frequency 0.6 THz (Fig. 4.5b). Another kind of meta-atoms with opposite chirality can be also designed by tuning the parameters of the DSPR. Finally, interleaving the two sets of DSPR with opposite chirality, and tailoring the orientation angles of those DSPR according two holographic images: the letter "L" and "R", multiplexed holography can be constructed with independent helicity channels. The similar chiral metasurfaces with asymmetric meta-atoms were also demonstrated in optical frequencies, as shown in Fig. 4.5d–f. Beyond the circular polarization multiplexing, the chiral metasurfaces can also be exploited to simultaneously realize asymmetric transmission and wavefront shaping [22].

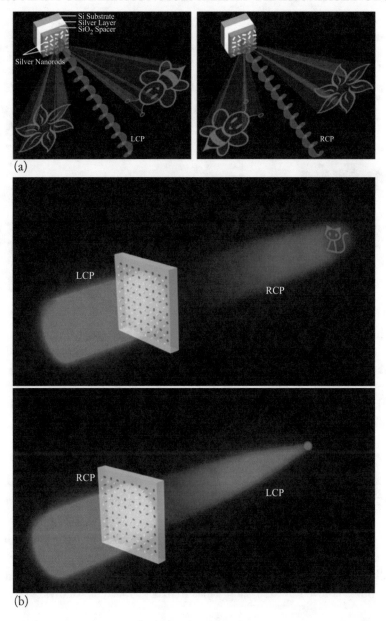

Figure 4.4: Helicity-dependent holographic imaging by geometric metasurfaces. (a) By interleaving two sets of meta-atoms to record two holographic images in the geometric metasurfaces. At the illumination of circular polarized incident light with different helicity, different images will appear at a specific position [17]. (b) Multiple functionalities of both holographic imaging and focusing realized by such helicity-multiplexed geometric metasurfaces [19].

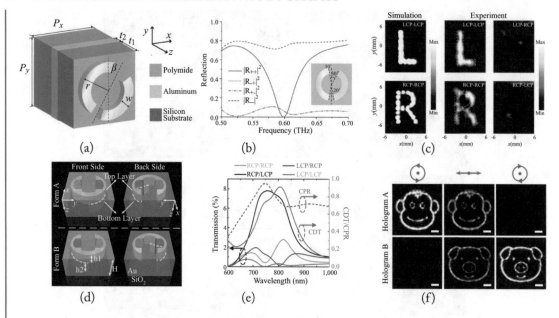

Figure 4.5: Chiral metasurface holograms that independently recording two holographic images by LCP and RCP polarizations. (a–c) The double-split ring resonators (DSPR) preform chiral response in its reflection spectrum at Terahertz wavelengths. By interleaving two types of DSPR with opposite chirality, multiplexed holographic dual images can be reconstructed independently [20]. (d–f) The similar chiral metasurface independently recording two holographic images are realized in the optical frequency range [21].

4.3   ARBITRARY POLARIZATION MULTIPLEXED METASURFACE HOLOGRAPHY

The geometric P-B phase provides an idea platform for circular polarization multiplexing, and the propagation phase provides high-performance platform for linear polarization multiplexing. By combing the geometric P-B phase and propagation phase, arbitrary elliptical polarization multiplexing can be achieved by analyzing Jones matrix [13, 23] as follows:

$$\mathbf{J} = \mathbf{R}^{-1}(\psi)\begin{pmatrix} e^{i\varphi_x} & 0 \\ 0 & e^{i\varphi_y} \end{pmatrix}\mathbf{R}(\psi), \tag{4.1}$$

where φ_x and φ_y are the propagation phase along two axis of the nanoposts, and ψ is the orientation angle that is related to the geometric P-B phase. Based on this Jones matrix, one can find the phase modulation rule on arbitrary orthogonal polarization states $\vec{\lambda}^+$ and $\vec{\lambda}^-$ as follows:

$$\mathbf{J}(x, y)\vec{\lambda}^+ = e^{i\phi^+(x,y)} \left(\vec{\lambda}^+\right)^*, \tag{4.2}$$

$$\mathbf{J}(x, y)\vec{\lambda}^- = e^{i\phi^-(x,y)} \left(\vec{\lambda}^-\right)^*, \tag{4.3}$$

where $*$ denotes the complex conjugate, $\phi^+(x, y)$ and $\phi^-(x, y)$ are two sets of independent phase profiles imparted on polarization sates $\vec{\lambda}^+$ and $\vec{\lambda}^-$, respectively. The two sets of phase profiles can record two independent holographic images as shown in Fig. 4.6a. Different from the previous chiral metasurfaces that independently record two holographic images by interleaving two different meta-atoms in a single unit cell, the independent dual holographic information is recorded in a single meta-atom in each unit cell, which has higher efficiency and better spatial resolution.

In addition to recording dual holographic images, the two sets of independent phase profiles can also record arbitrary special light beams. For example, azimuthal phase profiles with arbitrarily different topological charges m and n can be imparted on $\phi^+(x, y)$ and $\phi^-(x, y)$, respectively (Fig. 4.6b), forming a so-called J-plate [24–26]. Arbitrary spin to orbital conversion can be realized in such J-plate, which breaks the symmetry limitation of the conventional q-plate. The two independent phase profiles can also be employed to record two different accelerated light beams. For the LCP and RCP incident light, accelerated light beams with different curved trajectories will be generated (Fig. 4.6c). The arbitrary polarization manipulation by the metasurface were also used to demonstrate the quantum entanglement, in which the polarization states are represented by discrete quantum states of a single photon [27].

4.4   METASURFACE VECTORIAL HOLOGRAPHY

Polarization multiplexed meta-hologram only manipulates two orthogonal polarization states. Manipulating the polarization of light in an arbitrary way as shown in Fig. 4.7a, full vectorial properties of a light wavefront can be harnessed [29]. In the diatomic nanorod design, both the phase and polarization states are fully controlled by orientation angle and relative displacement parameters. By tailoring the global displacement of the diatomic meta-molecule, the phase profile that is responsible for the holographic image can be controlled. At the same time, the local displacement and orientation angle of the meta-molecule are tailored to modulate the spatial polarization state distribution of the holographic image. In this way, a vectorial meta-hologram is constructed, which can be used in anti-counterfeiting and data encryption applications. Based on the vectorial meta-hologram, a double way holographic switch can be built as shown in Fig. 4.7b. In the double-way holographic switch device, two sets of meta-molecules are interleaved in the metasurface design. The local displacement and orientation angle of the first/second set of meta-molecule are tailored to generate LHC/RHC from linear vertical polarized light. At the same time, the other degrees of freedom, the global displacements p_1, and p_2 of the meta-molecule

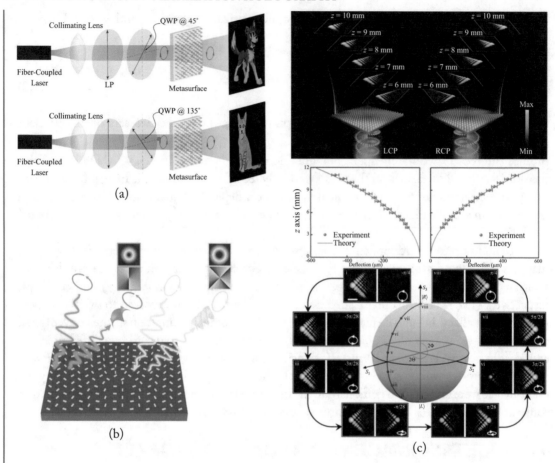

Figure 4.6: Arbitrary polarization multiplexing by metasurfaces that can fully control the phase and polarization states of light independently. An arbitrary polarization multiplexed metasurface can (a) record two independent holographic images by arbitrary elliptical polarization states [23], (b) realize arbitrary spin to orbital conversion [25], and demonstrate (c) spin-controlled arbitrary accelerating light beams [28].

are modulated to generate two sections of the *Taiji* holographic image. As a result, the brightness and darkness parts of the *Taiji* pattern have different polarization states. If we place a linear polarizer in front of the metasurface, and a circular polarizer in rear of the metasurface, flip either the linear polarizer or the circular polarizer while keeping the other polarizer unchanged can alternatively change the appearance of the *Taiji* pattern [29]. In addition to the plasmonic metasurfaces, all-dielectric metasurfaces can also realize such vectorial holograms [30–32].

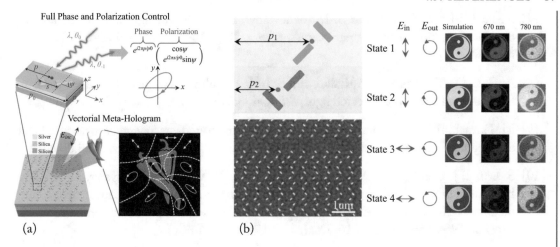

Figure 4.7: Vectorial meta-hologram that can realize (a) holographic images with arbitrarily distributed polarization states and (b) double-way holographic switches with arbitrary orthogonal polarization state pair [29].

4.5   REFERENCES

[1] J. W. Goodman, *Introduction to Fourier Optics*, Roberts and Company, 2004. DOI: 10.1063/1.3035549. 29

[2] L. Nikolova, and P. S. Ramanujam, *Polarization Holography*, Cambridge University Press, 2009. DOI: 10.1017/cbo9780511581489. 29

[3] T. Todorov, L. Nikolova, and N. Tomova, Polarization holography, vol. 1: A new high-efficiency organic material with reversible photoinduced birefringence, *Appl. Opt.*, 23(23):4309–4312, 1984. DOI: 10.1364/ao.23.004309. 29

[4] D. Ilieva, L. Nedelchev, Ts. Petrova, N. Tomova, V. Dragostinova, and L. Nikolova, Holographic multiplexing using photoinduced anisotropy and surface relief in azopolymer films, *J. Opt. A: Pure Appl. Opt.*, 7(1):35, 2005. DOI: 10.1088/1464-4258/7/1/005.

[5] T. Todorov, L. Nikolova, K. Stoyanova, and N. Tomova, Polarization holography. vol. 3: Some applications of polarization holographic recording, *Appl. Opt.*, 24(6):785–788, 1985. DOI: 10.1364/ao.24.000785. 29

[6] S. Boroviks, R. A. Deshpande, N. A. Mortensen, and S. I. Bozhevolnyi, Multifunctional metamirror: Polarization splitting and focusing, *ACS Photon.*, 5(5):1648–1653, 2018. DOI: 10.1021/acsphotonics.7b01091. 29

[7] A. Pors, M. G. Nielsen, and S. I. Bozhevolnyi, Plasmonic metagratings for simultaneous determination of Stokes parameters, *Optica*, 2(8):716–723, 2015. DOI: 10.1364/optica.2.000716.

[8] A. Pors, O. Albrektsen, I. P. Radko, and S. I. Bozhevolnyi, Gap plasmon-based metasurfaces for total control of reflected light, *Sci. Rep.*, 3:2155, 2013. DOI: 10.1038/srep02155.

[9] M. Farmahini-Farahani and H. Mosallaei, Birefringent reflectarray metasurface for beam engineering in infrared, *Opt. Lett.*, 38(4):462–464, 2013. DOI: 10.1364/ol.38.000462.

[10] S. Liu, T. J. Cui, Q. Xu, D. Bao, L. Du, X. Wan, W. X. Tang, C. Ouyang, X. Y. Zhou, H. Yuan, H. F. Ma, W. X. Jiang, J. Han, W. Zhang, and Q. Cheng, Anisotropic coding metamaterials and their powerful manipulation of differently polarized terahertz waves, *Light Sci. Appl.*, 5:e16076, 2016. DOI: 10.1038/lsa.2016.76. 29

[11] W. T. Chen, K.-Y. Yang, C.-M. Wang, Y.-W. Huang, G. Sun, I. D. Chiang, C. Y. Liao, W.-L. Hsu, H. T. Lin, S. Sun, L. Zhou, A. Q. Liu, and D. P. Tsai, High-efficiency broadband meta-hologram with polarization-controlled dual images, *Nano Lett.*, 14(1):225–230, 2013. DOI: 10.1021/nl403811d. 30

[12] Y. Montelongo, J. O. Tenorio-Pearl, W. I. Milne, and T. D. Wilkinson, Polarization switchable diffraction based on subwavelength plasmonic nanoantennas, *Nano Lett.*, 14(1):294–298, 2014. DOI: 10.1021/nl4039967. 30

[13] A. Arbabi, Y. Horie, M. Bagheri, and A. Faraon, Dielectric metasurfaces for complete control of phase and polarization with subwavelength spatial resolution and high transmission, *Nat. Nanotechnol.*, 10(11):937–943, 2015. DOI: 10.1038/nnano.2015.186. 30, 31, 34

[14] C. Min, J. Liu, T. Lei, G. Si, Z. Xie, J. Lin, L. Du, and X. Yuan, Plasmonic nano-slits assisted polarization selective detour phase meta-hologram, *Laser Photon. Rev.*, 10(6):978–985, 2016. DOI: 10.1002/lpor.201600101. 31

[15] Z.-L. Deng, S. Zhang, and G. P. Wang, A facile grating approach towards broadband, wide-angle and high-efficiency holographic metasurfaces, *Nanoscale*, 8(3):1588–1594, 2016. DOI: 10.1039/c5nr07181j. 31

[16] Z.-L. Deng, J. Deng, X. Zhuang, S. Wang, T. Shi, G. P. Wang, Y. Wang, J. Xu, Y. Cao, X. Wang, X. Cheng, G. Li, and X. Li, Facile metagrating holograms with broadband and extreme angle tolerance, *Light Sci. Appl.*, 7(1):78, 2018. DOI: 10.1038/s41377-018-0075-0. 31, 32

[17] D. Wen, F. Yue, G. Li, G. Zheng, K. Chan, S. Chen, M. Chen, K. F. Li, P. W. H. Wong, K. W. Cheah, E. Yue Bun Pun, S. Zhang, and X. Chen, Helicity multiplexed broadband

metasurface holograms, *Nat. Commun.*, 6:8241, 2015. DOI: 10.1038/ncomms9241. 32, 33

[18] M. Khorasaninejad, A. Ambrosio, P. Kanhaiya, and F. Capasso, Broadband and chiral binary dielectric meta-holograms, *Sci. Adv.*, 2(5):e1501258, 2016. DOI: 10.1126/sciadv.1501258. 32

[19] D. Wen, S. Chen, F. Yue, K. Chan, M. Chen, M. Ardron, K. F. Li, P. W. H. Wong, K. W. Cheah, E. Y. B. Pun, G. Li, S. Zhang, and X. Chen, Metasurface device with helicity-dependent functionality, *Adv. Opt. Mater.*, 4(2):321–327, 2016. DOI: 10.1002/adom.201500498. 32, 33

[20] Q. Wang, E. Plum, Q. Yang, X. Zhang, Q. Xu, Y. Xu, J. Han, and W. Zhang, Reflective chiral meta-holography: Multiplexing holograms for circularly polarized waves, *Light Sci. Appl.*, 7(1):25, 2018. DOI: 10.1038/s41377-018-0019-8. 32, 34

[21] F. Zhang, M. Pu, X. Li, P. Gao, X. Ma, J. Luo, H. Yu, and X. Luo, All-dielectric metasurfaces for simultaneous giant circular asymmetric transmission and wavefront shaping based on asymmetric photonic spin—orbit interactions, *Adv. Fun. Mater.*, 27(47):1704295, 2017. DOI: 10.1002/adfm.201770280. 32, 34

[22] Y. Chen, X. Yang, and J. Gao, Spin-controlled wavefront shaping with plasmonic chiral geometric metasurfaces, *Light Sci. and Appl.*, 7(1):84, 2018. DOI: 10.1038/s41377-018-0086-x. 32

[23] J. P. Balthasar Mueller, N. A. Rubin, R. C. Devlin, B. Groever, and F. Capasso, Metasurface polarization optics: Independent phase control of arbitrary orthogonal states of polarization, *Phys. Rev. Lett.*, 118(11):113901, 2017. DOI: 10.1103/physrevlett.118.113901. 34, 36

[24] R. C. Devlin, A. Ambrosio, N. A. Rubin, J. P. B. Mueller, and F. Capasso, Arbitrary spin-to-orbital angular momentum conversion of light, *Science*, 358:896–901, 2017. DOI: 10.1126/science.aao5392. 35

[25] S. Wang, F. Li, J. Deng, X. Ye, Z.-L. Deng, Y. Cao, B.-O. Guan, G. Li, and X. Li, Diatomic metasurface based broadband J-plate for arbitrary spin-to-orbital conversion, *J. Phys. D: Appl. Phys.*, 52:324002, 2019. DOI: 10.1088/1361-6463/ab2229. 36

[26] Y.-W. Huang, N. A. Rubin, A. Ambrosio, Z. Shi, R. C. Devlin, C.-W. Qiu, and F. Capasso, Versatile total angular momentum generation using cascaded J-plates, *Opt. Express*, 27(5):7469–7484, 2019. DOI: 10.1364/oe.27.007469. 35

[27] K. Wang, J. G. Titchener, S. S. Kruk, L. Xu, H.-P. Chung, M. Parry, I. I. Kravchenko, Y.-H. Chen, A. S. Solntsev, Y. S. Kivshar, D. N. Neshev, and A. A. Sukhorukov,

Quantum metasurface for multiphoton interference and state reconstruction, *Science*, 361(6407):1104–1108, 2018. DOI: 10.1126/science.aat8196. 35

[28] Q. Fan, W. Zhu, Y. Liang, P. Huo, C. Zhang, A. Agrawal, K. Huang, X. Luo, Y. Lu, C. Qiu, H. J. Lezec, and T. Xu, Broadband generation of photonic spin-controlled arbitrary accelerating light beams in the visible, *Nano Lett.*, 19(2):1158–1165, 2019. DOI: 10.1021/acs.nanolett.8b04571. 36

[29] Z.-L. Deng, J. Deng, X. Zhuang, S. Wang, K. Li, Y. Wang, Y. Chi, X. Ye, J. Xu, G. P. Wang, R. Zhao, X. Wang, Y. Cao, X. Cheng, G. Li, and X. Li, Diatomic metasurface for vectorial holography, *Nano Lett.*, 18(5):2885–2892, 2018. DOI: 10.1021/acs.nanolett.8b00047. 35, 36, 37

[30] R. Zhao, B. Sain, Q. Wei, C. Tang, X. Li, T. Weiss, L. Huang, Y. Wang, and T. Zentgraf, Multichannel vectorial holographic display and encryption, *Light Sci. Appl.*, 7(1):95, 2018. DOI: 10.1038/s41377-018-0091-0. 36

[31] E. Arbabi, S. M. Kamali, A. Arbabi, and A. Faraon, Vectorial holograms with a dielectric metasurface: Ultimate polarization pattern generation, *ACS Photon.*, 6(11):2712–2718, 2019. DOI: 10.1021/acsphotonics.9b00678.

[32] Y. Bao, J. Ni, and C.-W. Qiu, A minimalist single-layer metasurface for arbitrary and full control of vector vortex beams, *Adv. Mater.*, 1905659, 2019. DOI: 10.1002/adma.201905659. 36

CHAPTER 5

Metasurface Color Holography

Along with the polarization states of light, wavelength is another basic attribute of light. Manipulation of wavelength together with the metasurface is promising for many applications, such as optical dispersion-controlling devices [1–12], selective color routing [13, 14], and 3D full-color holographic displays [15–20].

Full-color holography has long been pursued for practical vivid 3D display applications. In earlier times, rainbow hologram has been used to construct colored holographic images [21, 22]. However, the color of the holographic image generated by rainbow hologram changes with the viewing angle, which largely prevents its practical applications. People also combined multiple spatial light modulators (SLMs) to construct the colorful holography [23–28], however, the optical setups are bulky and complicated, not to mention to other disadvantages including poor imaging quality, limited color gamut, and narrow viewing angle.

On the other hand, due to the flexibility of tuning the spectrum response of meta-atoms at will, color holography can be constructed in metasurface platform with high performances. There are two typical ways to realize metasurface wavelength multiplexing: (1) interleaving different sized meta-atoms that are resonant at different wavelengths to independently record different holograms [15–18]; and (2) non-interleaving approach with single-type meta-atoms by setting different incident angles for different wavelengths [19, 20].

5.1 INTERLEAVED COLOR MULTIPLEXING

In the metasurface platform, people first employ the resonant metasurfaces to realize the wavelength-multiplexed color holography. As shown in Fig. 5.1a, by tailoring the size of the aluminum nanorods, the resonance wavelength can be tuned to three wavelengths that correspond to the three primary colors: blue, green, and red [15]. For each set of nanorod, the sizes are further finely tuned to extract the resonance phase for holographic recording. Because both the required spectrum response and phase modulation are based on the resonance of the meta-atoms, the applicable size parameters of the nanorod are largely limited to simultaneously satisfy the requirement of spectrum response and phase modulation. In the example of Fig. 5.1a, only two phase levels were used to modulate the holographic images, in order to guarantee that the phase modulated nanorods are still near the resonant wavelength of the corresponding color. As a result, multiple high-order images always appear in the reconstructed plane, which degraded the diffraction efficiency of the color holography. Figure 5.1b shows a dual wavelength colored binary hologram by designing a diffractive unit with two kinds of plasmonic nanorods with dif-

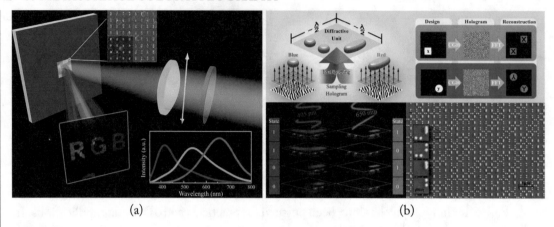

Figure 5.1: Color meta-holography by interleaving multiple sets of resonant meta-atoms. (a) Three sets of resonant nanorod blocks whose resonances occur at the three primary color wavelengths form a supercell pixel for full-color hologram [15]. (b) A single diffractive unit cell composed of multiple sized nanorods that response to different wavelengths. The present and absent states of the multiple sized nanorods represent binary phase modulations that are responsible for the holographic information carried on by different wavelengths [29].

ferent sizes [29]. The larger nanorod has a plasmonic resonance at 650 nm, and is employed to record the holographic image for red light. The smaller two nanorods resonate at shorter wavelength 405 nm, and are used to record another holographic image for blue light. Here, the double nanorods are used because the resonance amplitude for blue light is very weak, and double nanorods can strengthen the intensity of the diffracted blue light. In each diffraction unit, the absent and present states of the two kinds of nanorods represent two levels of phase modulations of the diffracted light. Because of this, twin images always exist and the diffraction efficiency is very low.

Another efficient way to construct multi-color meta-holograms can be implemented by interleaving multiple-sized meta-atoms with spatially varying orientations by exploiting the geometric P-B phase [16, 30]. As shown in Fig. 5.2a, three different-sized meta-atoms have different spectrum responses, whose peaks are located at the red (633 nm), green (532 nm), and blue (473 nm) wavelengths. As the spectrum peak value for blue light is much weaker than that for green and red color, one can arrange two blue meta-atoms together with one green and one red meta-atom in a single unit cell (Fig. 5.2b). In this way, each unit cell forms a compound pixel that can independently modulate the phases of three primary colored light. The phases of the three wavelengths are independently modulated by orientation angles of the three meta-atoms, which are completely decoupled from the frequency spectrum responses that are determined by the sizes of the meta-atoms. As a result, the modulated phase levels can be increased to eight,

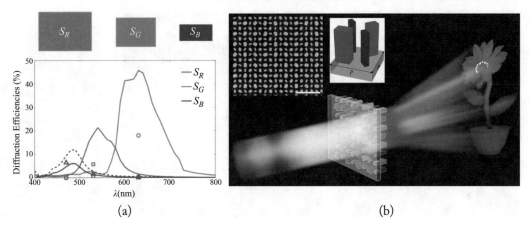

Figure 5.2: Color meta-holography by interleaving multiple sets of geometric P-B meta-atoms. (a) Thee different-sized meta-atoms have three different resonant peak diffraction efficiencies corresponding to the red, green, and blue colors. (b) Arranging the three sets of meta-atoms in a single unit cell, the orientation angles of different meta-atoms are independently tailored by the phase profile of different holographic information carried on by different wavelengths [16].

which is significantly improved compared with color-multiplexed resonant phase metasurfaces. Under the illumination of a circularly polarized white light beam composed of the three primary color light, both the conversion efficiency and image fidelity of the full-color holographic image were improved, as shown in Fig. 5.2b.

5.2   NON-INTERLEAVED COLOR MULTIPLEXING

Alternative to the interleaving approach to realize color multiplexing in metasurface holography, there is a non-interleaved approach by only using single type of meta-atoms. The non-interleaved color multiplexing is based on the introduction of different phase shifts by different wavelengths, which was previously used in traditional full-color holography construction [27] and ultrathin 2D material full-color holography [31]. In the hologram encoding process, if a phase shift term $\Delta\varphi$ is added, the holographic image deflects with an angle θ_0 with respect to the normal direction. The relation between the deflected angle and the phase shift term is

$$\frac{2\pi}{\lambda}\sin\theta_0 x = \Delta\varphi. \tag{5.1}$$

We can see that the deflection angle θ_0 is dependent on wavelength λ. It indicates that, for a given phase gradient term $\Delta\varphi$, the holographic images reconstructed by different colored light will be separated, as shown in the left panel of Fig. 5.3. To obtain a full-color image, three sets of different holographic information should be recorded by a single hologram, which means

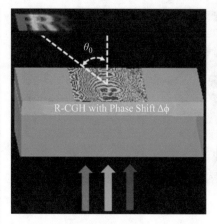  

Figure 5.3: Non-interleaved color-multiplexing by manipulating different incident angles of different wavelengths while employing single sets of meta-atoms. Adding appropriate phase shift terms by manipulating the incident angles of multiple wavelengths at a specific direction (e.g., x-direction), full-color hologram can be constructed [19].

that the reconstructed holographic image by different colored light should be different. One can impart phase shift terms with opposite signs for the red and blue components of the holographic information, respectively. At the same time, the incident red and blue beams are tilted with opposite angles, as shown in the middle panel of Fig. 5.3. In this way, in a specific observation area, the predesigned full-color holographic image could be precisely reconstructed. At the same time, there exists unwanted cross talk images with wrong color. However, the cross-talk images are spatially separated due to the imparted proper phase terms. The right panel of Fig. 5.3 shows the experimentally reconstructed full-color images that include not only the primary red, green, and blue colors, but also the secondary colors such as yellow, cyan, and white that come from combination of primary colors [19].

The similar non-interleaving color multiplexing can be demonstrated in a metasurface by making the phase shift term in both the x- and y-directions, and varying the incident angles of multiple beams in the k-space as demonstrated in [20]. If the phase gradient was large enough and the shift direction was chosen properly at the same time, all wrong-colored cross-talk images could be shifted out of the line cone, and only the correct full-color image was reconstructed in the free space. This strategy completely suppressed spatial cross-talk images in non-interleaving color-multiplexing, and is promising to improve the efficiency of full-color metasurface holography.

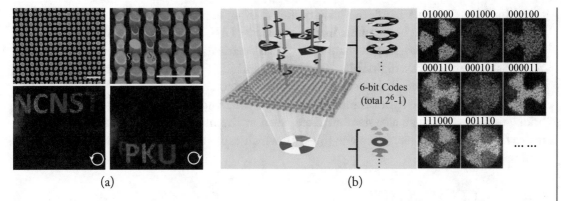

(a)   (b)

Figure 5.4: Hybrid polarization and color multiplexed metasurfaces. (a) Combining the interleaved color multiplexing and helicity multiplexing, polarization-dependent multiple full-color holographic images can be constructed [32]. (b) Combining the non-interleaved color multiplexing and helicity multiplexing, multiple channel digital states encoding can be realized [33].

5.3   HYBRID POLARIZATION AND COLOR MULTIPLEXING

By fully exploiting the degrees of freedom of the meta-atom, the full-color metasurface holography can be realized further by multiplexing in the polarization dimension. The hybrid polarization and color multiplexing can dramatically increase the multiplexing channels of recorded holographic information and largely extend the functionalities of the metasurface [32–35].

In Fig. 5.4a, the interleaving color-multiplexing geometric metasurface was combined with the helicity multiplexing of the geometric P-B phase. For the color-multiplexing, a four-element compound unit cell was employed to work as a super pixel to independently modulate the phase of the primary three colors. The sizes of the silicon nanopost that construct the metasurface determine the response wavelength of light, and the orientation angles determine the modulated phase when a circularly polarized light impinges on it. As discussed previously in Section 4.2, the geometric P-B phase metasurface can readily be designed for helicity multiplexing to generate different single-colored holographic images in a particular area for LHC and RHC illuminations. As the geometric P-B phase is irrespective with the wavelength, the color-multiplexed geometric phase can be realized further by multiplexing the helicity. As shown in lower panels of Fig. 5.4a, different full-color holographic image appears in the observation zone when the metasurface was illuminated by the LHC and RHC polarized light, respectively.

The hybrid polarization and color multiplexing can also be used for multiple digital bits encoding, as the multiplexing channels are significantly increased compared with pure polarization or wavelength multiplexing. As shown in Fig. 5.4b, a non-interleaving multiplexing of three different colored images at a given z-plane was produced by a multi-wavelength Gerchberg–Saxton algorithm. On the other hand, the helicity multiplexing approach can be simultaneously employed to focus different holographic images carried by LHC and RHC in the same plane. The combination of three wavelength channels and two polarization channels gave rise to a six base channels that represent six digital bits. Based on the six bits, $2^6 - 1$ units of different holographic information could be encoded by such a single metasurface.

In addition to the hybridization of color and polarization multiplexing, many other hybridization schemes were also demonstrated, such as the mixed hologram and color printing realized in a single metasurface plane [36–39], which exploited the full potential of the metasurface holography.

5.4   REFERENCES

[1] F. Aieta, M. A. Kats, P. Genevet, and F. Capasso, Multiwavelength achromatic metasurfaces by dispersive phase compensation, *Science*, 347(6228):1342–1345, 2015. DOI: 10.1126/science.aaa2494. 41

[2] Z.-L. Deng, S. Zhang, and G. P. Wang, Wide-angled off-axis achromatic metasurfaces for visible light, *Opt. Express*, 24(20):23118–23128, 2016. DOI: 10.1364/oe.24.023118.

[3] M. Khorasaninejad, F. Aieta, P. Kanhaiya, M. A. Kats, P. Genevet, D. Rousso, and F. Capasso, Achromatic metasurface lens at telecommunication wavelengths, *Nano Lett.*, 15(8):5358–5362, 2015. DOI: 10.1021/acs.nanolett.5b01727.

[4] Z. Zhao, M. Pu, H. Gao, J. Jin, X. Li, X. Ma, Y. Wang, P. Gao, and X. Luo, Multispectral optical metasurfaces enabled by achromatic phase transition, *Sci. Rep.*, 5:15781, 2015. DOI: 10.1038/srep15781.

[5] M. Pu, Z. Zhao, Y. Wang, X. Li, X. Ma, C. Hu, C. Wang, C. Huang, and X. Luo, Spatially and spectrally engineered spin-orbit interaction for achromatic virtual shaping, *Sci. Rep.*, 5:9822, 2015. DOI: 10.1038/srep09822.

[6] Y. Li, X. Li, M. Pu, Z. Zhao, X. Ma, Y. Wang, and X. Luo, Achromatic flat optical components via compensation between structure and material dispersions, *Sci. Rep.*, 6:19885, 2016. DOI: 10.1038/srep19885.

[7] M. Pu, X. Li, X. Ma, Y. Wang, Z. Zhao, C. Wang, C. Hu, P. Gao, C. Huang, H. Ren, X. Li, F. Qin, J. Yang, M. Gu, M. Hong, and X. Luo, Catenary optics for achromatic generation of perfect optical angular momentum, *Sci. Adv.*, 1(9), 2015. DOI: 10.1126/sciadv.1500396.

[8] J. Cheng and H. Mosallaei, Truly achromatic optical metasurfaces: A filter circuit theory-based design, *J. Opt. Soc. Am. B*, 32(10):2115–2121, 2015. DOI: 10.1364/josab.32.002115.

[9] S. Wang, J. Lai, T. Wu, C. Chen, and J. Sun, Wide-band achromatic flat focusing lens based on all-dielectric subwavelength metasurface, *Opt. Express*, 25(6):7121–7130, 2017. DOI: 10.1364/oe.25.007121.

[10] O. Avayu, E. Almeida, Y. Prior, and T. Ellenbogen, Composite functional meta-surfaces for multispectral achromatic optics, *Nat. Commun.*, 8:14992, 2017. DOI: 10.1038/ncomms14992.

[11] K. Li, Y. Guo, M. Pu, X. Li, X. Ma, Z. Zhao, and X. Luo, Dispersion controlling meta-lens at visible frequency, *Opt. Express*, 25(18):21419–21427, 2017. DOI: 10.1364/oe.25.021419.

[12] M. Khorasaninejad, Z. Shi, A. Y. Zhu, W. T. Chen, V. Sanjeev, A. Zaidi, and F. Capasso, Achromatic metalens over: 60 nm bandwidth in the visible and metalens with reverse chromatic dispersion, *Nano Lett.*, 17(3):1819–1824, 2017. DOI: 10.1021/acs.nanolett.6b05137. 41

[13] C. Yan, K.-Y. Yang, and O. J. F. Martin, Fano-resonance-assisted metasurface for color routing, *Light Sci. Appl.*, 6:e17017, 2017. DOI: 10.1038/lsa.2017.17. 41

[14] B. H. Chen, P. C. Wu, V.-C. Su, Y.-C. Lai, C. H. Chu, I. C. Lee, J.-W. Chen, Y. H. Chen, Y.-C. Lan, C.-H. Kuan, and D. P. Tsai, GaN metalens for pixel-level full-color routing at visible light, *Nano Lett.*, 2017. DOI: 10.1021/acs.nanolett.7b03135. 41

[15] Y.-W. Huang, W. T. Chen, W.-Y. Tsai, P. C. Wu, C.-M. Wang, G. Sun, and D. P. Tsai, Aluminum plasmonic multicolor meta-hologram, *Nano Lett.*, 15(5):3122–3127, 2015. DOI: 10.1021/acs.nanolett.5b00184. 41, 42

[16] B. Wang, F. Dong, Q.-T. Li, D. Yang, C. Sun, J. Chen, Z. Song, L. Xu, W. Chu, Y.-F. Xiao, Q. Gong, and Y. Li, Visible-frequency dielectric metasurfaces for multiwavelength achromatic and highly dispersive holograms, *Nano Lett.*, 16(8):5235–5240, 2016. DOI: 10.1021/acs.nanolett.6b02326. 42, 43

[17] W. Zhao, B. Liu, H. Jiang, J. Song, Y. Pei, and Y. Jiang, Full-color hologram using spatial multiplexing of dielectric metasurface, *Opt. Lett.*, 41(1):147–150, 2016. DOI: 10.1364/ol.41.000147.

[18] S. Choudhury, U. Guler, A. Shaltout, V. M. Shalaev, A. V. Kildishev, and A. Boltasseva, Pancharatnam—berry phase manipulating metasurface for visible color hologram based on low loss silver thin film, *Adv. Opt. Mater.*, 5(10):1700196, 2017. DOI: 10.1002/adom.201700196. 41

[19] W. Wan, J. Gao, and X. Yang, Full-color plasmonic metasurface holograms, *ACS Nano*, 10(12):10671–10680, 2016. DOI: 10.1021/acsnano.6b05453. 41, 44

[20] X. Li, L. Chen, Y. Li, X. Zhang, M. Pu, Z. Zhao, X. Ma, Y. Wang, M. Hong, and X. Luo, Multicolor 3D meta-holography by broadband plasmonic modulation, *Sci. Adv.*, 2(11):e1601102, 2016. DOI: 10.1126/sciadv.1601102. 41, 44

[21] J. W. Goodman, *Introduction to Fourier Optics*, Roberts and Company, 2004. DOI: 10.1063/1.3035549. 41

[22] T. C. Poon, *Digital Holography and Three-Dimensional Display*, Springer, 2006. DOI: 10.1007/0-387-31397-4. 41

[23] D. Alfieri, G. Coppola, S. De Nicola, P. Ferraro, A. Finizio, G. Pierattini, and B. Javidi, Method for superposing reconstructed images from digital holograms of the same object recorded at different distance and wavelength, *Opt. Commun.*, 260(1):113–116, 2006. DOI: 10.1016/j.optcom.2005.10.055. 41

[24] S. Yeom, B. Javidi, P. Ferraro, D. Alfieri, S. DeNicola, and A. Finizio, Three-dimensional color object visualization and recognition using multi-wavelength computational holography, *Opt. Express*, 15(15):9394–9402, 2007. DOI: 10.1364/oe.15.009394.

[25] T. Wang, Y. Yu, and H. Zheng, Method for removing longitudinal chromatism in full color holographic projection system, *Opt. Eng.*, 50(9):091302–091305, 2011. DOI: 10.1117/1.3596175.

[26] T. Kozacki and M. Chlipala, Color holographic display with white light LED source and single phase only SLM, *Opt. Express*, 24(3):2189–2199, 2016. DOI: 10.1364/oe.24.002189.

[27] S.-F. Lin and E.-S. Kim, Single SLM full-color holographic: 3D display based on sampling and selective frequency-filtering methods, *Opt. Express*, 25(10):11389–11404, 2017. DOI: 10.1364/oe.25.011389. 43

[28] Z. Zeng, H. Zheng, Y. Yu, A. K. Asundi, and S. Valyukh, Full-color holographic display with increased-viewing-angle [Invited], *Appl. Opt.*, 56(13):F112–F120, 2017. DOI: 10.1364/ao.56.00f112. 41

[29] Y. Montelongo, J. O. Tenorio-Pearl, C. Williams, S. Zhang, W. I. Milne, and T. D. Wilkinson, Plasmonic nanoparticle scattering for color holograms, *PNAS*, 111(35):12679–12683, 2014. DOI: 10.1073/pnas.1405262111. 42

[30] B. Wang, F. Dong, D. Yang, Z. Song, L. Xu, W. Chu, Q. Gong, and Y. Li, Polarization-controlled color-tunable holograms with dielectric metasurfaces, *Optica*, 4(11):1368–1371, 2017. DOI: 10.1364/optica.4.001368. 42

[31] X. Li, H. Ren, X. Chen, J. Liu, Q. Li, C. Li, G. Xue, J. Jia, L. Cao, A. Sahu, B. Hu, Y. Wang, G. Jin, and M. Gu, Athermally photoreduced graphene oxides for three-dimensional holographic images, *Nat. Commun.*, 6:6984, 2015. DOI: 10.1038/ncomms7984. 43

[32] F. Dong, H. Feng, L. Xu, B. Wang, Z. Song, X. Zhang, L. Yan, X. Li, Y. Tian, W. Wang, L. Sun, Y. Li, and W. Chu, Information encoding with optical dielectric metasurface via independent multichannels, *ACS Photon.*, 6(1):230–237, 2019. DOI: 10.1021/acsphotonics.8b01513. 45

[33] L. Jin, Z. Dong, S. Mei, Y. F. Yu, Z. Wei, Z. Pan, S. D. Rezaei, X. Li, A. I. Kuznetsov, Y. S. Kivshar, J. K. W. Yang, and C.-W. Qiu, Noninterleaved metasurface for, (26-1) spin- and wavelength-encoded holograms, *Nano Lett.*, 18(12):8016–8024, 2018. DOI: 10.1021/acs.nanolett.8b04246. 45

[34] Y. Hu, L. Li, Y. Wang, M. Meng, L. Jin, X. Luo, Y. Chen, X. Li, S. Xiao, H. Wang, Y. Luo, C.-W. Qiu, and H. Duan, Trichromatic and tripolarization-channel holography with noninterleaved dielectric metasurface, *Nano Lett.*, 20(2):994–1002, 2020. DOI: 10.1021/acs.nanolett.9b04107.

[35] Z.-L. Deng, M. Jin, X. Ye, S. Wang, T. Shi, J. Deng, N. Mao, Y. Cao, B.-O. Guan, A. Alù, G. Li, and X. Li, Full-color complex-amplitude vectorial holograms based on multifreedom metasurfaces, *ArXiv:1912.11184*, 2019. DOI: 10.1002/adfm.201910610. 45

[36] Y. Zhang, L. Shi, D. Hu, S. Chen, S. Xie, Y. Lu, Y. Cao, Z. Zhu, L. Jin, B.-O. Guan, S. Rogge, and X. Li, Full-visible multifunctional aluminium metasurfaces by in situ anisotropic thermoplasmonic laser printing, *Nanoscale Horiz.*, 4(3):601–609, 2019. DOI: 10.1039/c9nh00003h. 46

[37] Y. Hu, X. Luo, Y. Chen, Q. Liu, X. Li, Y. Wang, N. Liu, and H. Duan, 3D-integrated metasurfaces for full-colour holography, *Light Sci. Appl.*, 8(1):86, 2019. DOI: 10.1038/s41377-019-0198-y.

[38] Y. Bao, Y. Yu, H. Xu, C. Guo, J. Li, S. Sun, Z.-K. Zhou, C.-W. Qiu, and X.-H. Wang, Full-colour nanoprint-hologram synchronous metasurface with arbitrary hue-saturation-brightness control, *Light Sci. Appl.*, 8(1):95, 2019. DOI: 10.1038/s41377-019-0206-2.

[39] Q. Wei, B. Sain, Y. Wang, B. Reineke, X. Li, L. Huang, and T. Zentgraf, Simultaneous spectral and spatial modulation for color printing and holography using all-dielectric metasurfaces, *Nano Lett.*, 19(12):8964–8971, 2019. DOI: 10.1021/acs.nanolett.9b03957. 46

CHAPTER 6

Surface-Wave and Metagrating Holography

Traditional holography only deals with the wavefront shaping of free-space waves through interference of multiple coherent light beams. Holographic interference technique has been successfully applied in surface plasmon waves [1], and the wavefronts of both free-space wave and surface wave could be shaped by a general holographic way in a metagrating configuration. The free-space wave and surface wave could be mutually converted in different diffraction orders of the metagrating through the reciprocal lattice vector. It is possible to generate arbitrary wavefront shapes of both surface wave and free-space wave by making a properly designed metagrating. By modulating the metagrating according to the interference pattern of a free-space reference wave and an arbitrary surface object wave, any surface wavefront on demand could be generated by a free-space wave excitation. On the other hand, making a metagrating by the interference of a surface reference wave and arbitrary free-space object wave, any 3D free-space holographic images could be generated by a surface wave excitation in an integrated platform.

6.1 METAGRATINGS FOR SURFACE WAVEFRONT SHAPING

The wavefront shaping of a surface wave provides an ability to produce versatile surface plasmon profiles in a controllable manner, which is of important interest in the field of plasmonic lithography and imaging devices [2–8]. A surface plasmon plane wave can be excited by a free-space plane wave in high diffraction orders of a periodic metallic grating. The conversion of the free-space wave and the surface wave is based on the momentum exchange provided by the periodicity of the grating. As the excitation of surface plasmon wave is highly dependent on the polarization of the incident light and periodicity of the meta-atom, the orientation angle and displacement of the meta-atoms can be readily tailored to modulate both the amplitude and phase information of a surface plasmon wave. As shown in Fig. 6.1, the unit structure for surface wavefront shaping is composed of a pair of nanoslits perforated in a thin metallic film. If the double slits are placed to perpendicular to each other with a fixed center-to-center distance that is half of the surface plasmon wavelength $S = \lambda_{SP}/2$, the unit structure could work as an effective half waveplate for surface waves. The global orientation angle of the perpendicular double slits produces geometric phases across the full 2π range for the generated surface plas-

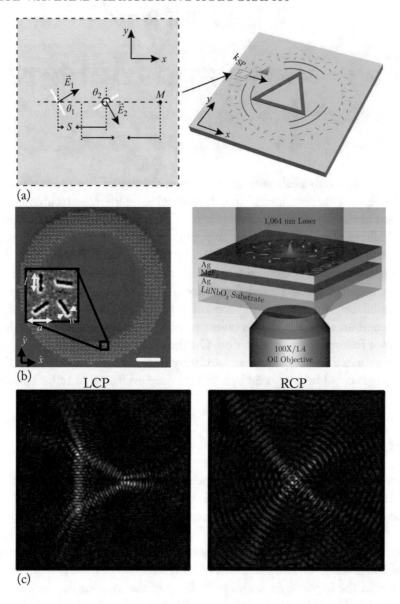

Figure 6.1: Surface plasmon wavefront shaping by tailoring the interference pattern of surface waves scattered from metallic slit arrays. (a) Schematic of the surface wave holographic image pattern generation by tuning the orientation angles of nanoslit pair arrays [2]. (b) The experimental realized surface plasmon meta-hologram and optical setup for surface wave holographic image reconstruction [3]. (c) The reconstructed surface wave holographic images by different circular polarized incident beams [3].

mon wave excited by circularly polarized free-space wave. To further freely tailor the orientation angle θ_1 and θ_2 of each nanoslit independently, both the amplitude and phase of the generated surface plasmon wave could be modulated through the orientation angle difference $(\theta_1 - \theta_2)$, and the orientation angle summation $(\theta_1 + \theta_2)$, respectively. As a result, any desired amplitude and phase profile of the surface plasmon wave could be generated by properly designing the two orientation angles θ_1 and θ_2. Based on this modulation principle, a surface wave hologram ring is designed as shown in the right panel of Fig. 6.1a, which can generate arbitrary 2D surface wave patterns in the central area surrounded by the hologram ring. Figure 6.1b shows the SEM image of the fabricated surface wave hologram ring and the experimental setup for reconstruction of the surface wave holographic image. The free-space light beam from a 1064 nm laser was coupled to the hologram ring by normal incidence. The predesigned holographic image was generated as the near field pattern formed by the surface plasmon wave, which was collected by a high numerical aperture objective lens at the bottom of the hologram ring.

The hologram ring also supports spin-dependent phenomena related to the optical spin-Hall effect, such as the spin-dependent splitting of beam displacement [9] and the spin-dependent directional propagation of surface plasmon waves [6, 10]. However, due to the time-reversal symmetry, the phase profiles related to two opposite spins are always related to each other by an opposite sign in previous optical spin-Hall effect. It means that the incident light with opposite spins can only generate simple and symmetric behaviors, which are far from arbitrary and independent manipulation for those two spins. The hologram ring here can independently control the surface plasmon orbitals for two opposite spins. The two freely controllable orientation angles of the nanoslit pair unit can be properly designed to generate different focal spots for two opposite spins, and thus give completely different near-field profiles of the surface plasmon waves, as shown in Fig. 6.1c.

6.2   METAGRATINGS FOR FREE-SPACE WAVEFRONT SHAPING

In addition to modulating arbitrary surface wavefront generated from a free space reference wave, inversely, the plasmonic metagrating can also modulate arbitrary free-space wavefront by an integrated surface wave excitation [11–16]. As shown in Fig. 6.2a, the left side part of the plasmonic metagrating works as a coupler from the free space wave to a SPP wave, which is a typical setup for surface plasmon wave generations. When the propagated surface plasmon wave meets the right part of the metagrating, it will be decoupled from the surface and radiate back to the free space along another diffraction order of the metagrating. The principle is based on periodic metagratings that deal with plane free-space wave and plane surface wave; however, it is applicable to arbitrary wavefront conversion between surface wave and propagation wave if the period of the metagrating is properly modulated. It means that, when the right part metagrating of Fig. 6.2a is modulated by a predesigned phase profile, for example, cubic phase profile as

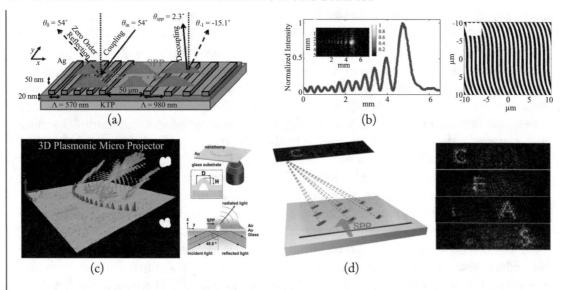

(a)   (b)

(c)   (d)

Figure 6.2: Metagratings that convert surface waves into arbitrary free-space holographic wavefront. (a) Schematic of the plasmonic metagrating that connect the SPP waves in free-space waves through diffraction orders of the grating [11]. (b) Modulated metagratings for free-space Airy beam generations by surface plasmon wave excitation [11]. (c) Modulated metallic bump metagrating that works as a 3D plasmonic micro projector [12]. (d) Modulated metallic slit metagratings that can convert SPP waves with different polarizations and propagation directions into different holographic images in free-space [13].

shown in Fig. 6.2b, the decoupled free-space wave from the surface plasmon wave could be the desired Airy beam.

Such wavefront shaping techniques can be served as a 3D plasmonic micro projector in an integrated platform without the bulky light source illumination (Fig. 6.2c). The input light source was integrated in the device as the surface plasmon wave propagating along the metallic surface; the desired 3D free-space holographic image was directly generated by the surface structure with the help of a series of nano bumps at the metallic surface. The texture of the nanobump arrays is properly arranged as the interferogram between the SPP reference wave and the objected wavefront emitted by the projected 3D holographic image.

Exploiting the anisotropy of the meta-atoms that forms the metagrating, the polarization states of light can be simultaneously controlled by surface plasmon wave sources. As shown in Fig. 6.2d, by designing the metallic slits unit with properly tailored orientation angle and relative displacement, multiple free-space holographic images with different polarization sates could be constructed. By incorporating the SPP propagation direction and the polarization-selective

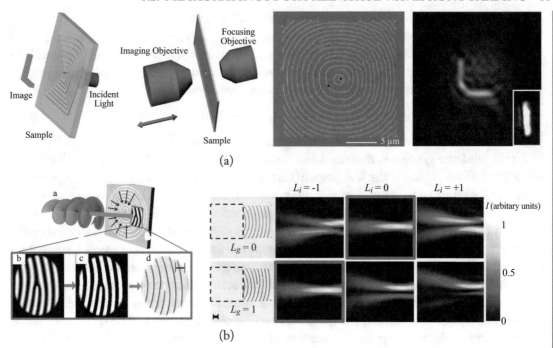

Figure 6.3: Modulated circular metagratings that convert cylindrical SPP wave into arbitrary free-space wavefront. (a) A circular metallic groove metagrating with a perforated central hole. Cylindrical SPP wave excited through the hole interference with the modulated metallic grooves, which give rise to predesigned holographic image in the far field [14]. (b) A circular metallic slit metagrating that is formed by the interference pattern of a SPP wave and a free-space vortex beam, which can be used as a holographic detector of the orbital angular momentum of light with plasmonic photodiodes [15].

scattering, the multiplexing channels are increased to four, breaking the channel number limit of traditional polarization multiplexing.

In addition to above straight metagratings that deal with reference planar surface wave, another typical way for shaping free-space wavefront from surface wave could be implemented in circular plasmonic metagratings that deal with reference cylindrical surface wave by a point source. As shown in Fig. 6.3a, the metagrating is composed of a thin metallic film drilled with a perforated hole and a series of circular grooves surrounding the hole. A cylindrical surface plasmon wave could be readily excited by focusing the input free-space light beam through an objective lens on the back of the metallic film [14]. On the front side of the metallic film, cylindrical surface plasmon waves were scattered out of the free space by those grooves. The positions of the circular grooves were carefully designed so that a Latin letter "L" could be scattered in the free-space above the metallic grooves. The phase distribution of the cylindrical surface plasmon wave is analytically written as $\varphi(x, y) = n_{spp}k_0 r$, $r = \sqrt{x^2 + y^2}$, and the phase profile of the holographic image $\varphi_{obj}(x, y)$ was calculated by the diffraction field emitted by a source array that has the "L" pattern. The position of the circular grooves can be determined by the constructive interference condition between those two waves, that is,

$$\varphi_{obj}(x, y) + n_{spp}k_0 r = 2m\pi, \tag{6.1}$$

where m is an integer. The modulated circular grooves are as shown in the third panel of Fig. 6.3a. The texture of the grooves is deformed and some of them are even discontinuous due to the modulated phase profile of the holographic image. The "L" shaped holographic image can be readily collected by an imaging objective, as shown in the fourth panel of Fig. 6.3a.

The circular metagrating can further be integrated with electric devices to build a holographic photo-detector. Figure 6.3b shows a metallic groove shaped device that is formed by the interference pattern of a cylindrical surface plasmon wave and a free-space vortex beam with a given orbital angular momentum (OAM). When a vortex beam with the predesigned angular momentum illuminate on the structure, a cylindrical surface plasmon wave could be generated in the metallic surface and be focused into a specific area. However, if the incident beam carries on another angular momentum, the generated surface plasmon wave will be defocused due to the mismatch of the total angular momentums of the whole system. The OAM-selective focusing of the surface plasmon wave can be used for sensitive OAM detection. The OAM detector is achieved by making scattering patterns in the focused area of the surface plasmon wave, and connect the scattering pattern to a photodetector [15]. The incident beam with the correct OAM will generated focused surface plasmon wave, whose energy will be collected by the photodetector, while the incident beams carrying any other OAM will be defocused from the scattering pattern and thus will not be detected.

The metagratings platform builds a solid bridge between the wavefront shaping of surface plasmon wave and free-space wave. In addition, the surface plasmon wave, in both its propagating and localized formalities, can also be employed to perfectly channel light between desired diffraction orders of the metagrating [17–20]. As shown in Fig. 6.4a, at the excitation of propagating or localized surface plasmon mode in a metallic metagrating, the incident free-space light beam can be perfectly channeled to the −1st diffraction order, which was refer to as the extraordinary optical diffraction (EOD). The EOD occurs in a particular regime surrounded by the −1st, 1st, and −2nd Rayleigh anomaly curves (Fig. 6.4b), where only the 0th and −1st free-space diffraction orders were supported by the metagrating. At this EOD regime, the surface plasmon resonance can lead to complete suppression of specular reflection and the perfect diffraction in the −1st diffraction order. Based on such perfect diffraction metagrating, arbitrary free-space wavefront can be high-efficiently shaped by modulating fringe pattern according to a given phase profile. As shown in Fig. 6.4c, setting the fringe pattern of the EOD metagrating as the interferogram between a spherical wave and a plane wave, the metagrating can completely focus an incident free-space light beam at its −1st diffraction order, and with the total sup-

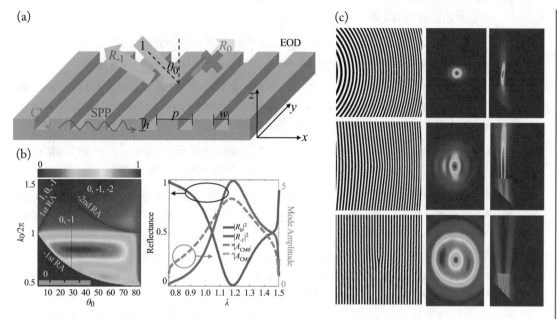

Figure 6.4: Metagratings that covert free-space plane-waves into arbitrary free-space wavefront with near-unitary diffraction efficiency. (a) Schematic of the metagrating that totally funnel incident light into −1st diffraction orders under the excitation of cavity mode in the metallic slits or the SPP wave, which was refer to as the extraordinary optical diffraction (EOD). (b) Diffraction efficiencies for varying incident angles θ_0 and wavenumbers k_0. The unitary diffraction peak corresponding to the mode amplitude peak of the cavity mode. (c) The modulated EOD metagratings that can highly efficiently convert plane waves into focused beam, Bessel beam, and vortex beam, etc. [17].

pression in the 0th diffraction order (upper panel in Fig. 6.4c). Similarly, modulating the EOD metagrating as the interference pattern between a vortex beam and a plane wave beam forms a fork-shaped structure, which can perfectly convert an incident Gaussian beam into a vortex beam in the −1st diffraction order, and the unwanted zero-order diffraction is also completely suppressed (middle panel of Fig. 6.4c). The metagrating formed by the interference pattern of a plane wave and a Bessel beam works as a perfect axicon that can focus a Gaussian into a finite line segment, which is a signature of a practical Bessel-Gaussian beam (lower panel of Fig. 6.4c).

Beyond the plasmonic metagratings, there also emerge on-chip wavefront shaping with dielectric metasurfaces recently [21–24], which subtly bridges the integrated waveguide modes and free-space wavefront.

6.3 REFERENCES

[1] M. Ozaki, J.-i. Kato, and S. Kawata, Surface-plasmon holography with white-light illumination, *Science*, 332(6026):218–220, 2011. DOI: 10.1126/science.1201045. 51

[2] Q. Xu, X. Zhang, Y. Xu, C. Ouyang, Z. Tian, J. Gu, J. Li, S. Zhang, J. Han, and W. Zhang, Polarization-controlled surface plasmon holography, *Laser Photon. Rev.*, 1600212–n/a, 2016. DOI: 10.1002/lpor.201770002. 51, 52

[3] S. Xiao, F. Zhong, H. Liu, S. Zhu, and J. Li, Flexible coherent control of plasmonic spin-Hall effect, *Nat. Commun.*, 6, 2015. DOI: 10.1038/ncomms9360. 52

[4] Y.-G. Chen, F.-Y. Yang, J. Liu, and Z.-Y. Li, Broadband focusing and demultiplexing of surface plasmon polaritons on metal surface by holographic groove patterns, *Opt. Express*, 22(12):14727–14737, 2014. DOI: 10.1364/oe.22.014727.

[5] Q. Tan, Q. Guo, H. Liu, X. Huang, and S. Zhang, Controlling the plasmonic orbital angular momentum by combining the geometric and dynamic phases, *Nanoscale*, 9(15):4944–4949, 2017. DOI: 10.1039/c7nr00124j.

[6] J. Lin, J. P. B. Mueller, Q. Wang, G. Yuan, N. Antoniou, X.-C. Yuan, and F. Capasso, Polarization-controlled tunable directional coupling of surface plasmon polaritons, *Science*, 340(6130):331–334, 2013. DOI: 10.1126/science.1233746. 53

[7] X. M. Tang, L. Li, T. Li, Q. J. Wang, X. J. Zhang, S. N. Zhu, and Y. Y. Zhu, Converting surface plasmon to spatial airy beam by graded grating on metal surface, *Opt. Lett.*, 38(10):1733–1735, 2013. DOI: 10.1364/ol.38.001733.

[8] L. Li, T. Li, S. M. Wang, C. Zhang, and S. N. Zhu, Plasmonic airy beam generated by in-plane diffraction, *Phys. Rev. Lett.*, 107(12):126804, 2011. DOI: 10.1103/physrevlett.107.126804. 51

[9] X. Yin, Z. Ye, J. Rho, Y. Wang, and X. Zhang, Photonic spin hall effect at metasurfaces, *Science*, 339(6126):1405–1407, 2013. DOI: 10.1126/science.1231758. 53

[10] L. Huang, X. Chen, B. Bai, Q. Tan, G. Jin, T. Zentgraf, and S. Zhang, Helicity dependent directional surface plasmon polariton excitation using a metasurface with interfacial phase discontinuity, *Light Sci. Appl.*, 2:e70, 2013. DOI: 10.1038/lsa.2013.26. 53

[11] I. Dolev, I. Epstein, and A. Arie, Surface-plasmon holographic beam shaping, *Phys. Rev. Lett.*, 109(20):203903, 2012. DOI: 10.1103/physrevlett.109.203903. 53, 54

[12] C. M. Chang, M. L. Tseng, B. H. Cheng, C. H. Chu, Y. Z. Ho, H. W. Huang, Y.-C. Lan, D.-W. Huang, A. Q. Liu, and D. P. Tsai, Three-dimensional plasmonic micro projector for light manipulation, *Adv. Mater.*, 25(8):1118–1123, 2013. DOI: 10.1002/adma.201203308. 54

[13] J. Chen, T. Li, S. Wang, and S. Zhu, Multiplexed holograms by surface plasmon propagation and polarized scattering, *Nano Lett.*, 17(8):5051–5055, 2017. DOI: 10.1021/acs.nanolett.7b02295. 54

[14] Y.-H. Chen, L. Huang, L. Gan, and Z.-Y. Li, Wavefront shaping of infrared light through a subwavelength hole, *Light Sci. Appl.*, 1:e26, 2012. DOI: 10.1038/lsa.2012.26. 55

[15] P. Genevet, J. Lin, M. A. Kats, and F. Capasso, Holographic detection of the orbital angular momentum of light with plasmonic photodiodes, *Nat. Commun.*, 3:1278, 2012. DOI: 10.1038/ncomms2293. 55, 56

[16] L. Li, T. Li, X.-M. Tang, S.-M. Wang, Q.-J. Wang, and S.-N. Zhu, Plasmonic polarization generator in well-routed beaming, *Light Sci. Appl.*, 4:e330, 2015. DOI: 10.1038/lsa.2015.103. 53

[17] Z.-L. Deng, S. Zhang, and G. P. Wang, A facile grating approach towards broadband, wide-angle and high-efficiency holographic metasurfaces, *Nanoscale*, 8(3):1588–1594, 2016. DOI: 10.1039/c5nr07181j. 56, 57

[18] Z.-L. Deng, S. Zhang, and G. P. Wang, Wide-angled off-axis achromatic metasurfaces for visible light, *Opt. Express*, 24(20):23118–23128, 2016. DOI: 10.1364/oe.24.023118.

[19] Z.-L. Deng, Y. Cao, X. Li, and G. P. Wang, Multifunctional metasurface: From extraordinary optical transmission to extraordinary optical diffraction in a single structure, *Photon. Res.*, 6(5):443–450, 2018. DOI: 10.1364/prj.6.000659.

[20] Z.-L. Deng, J. Deng, X. Zhuang, S. Wang, T. Shi, G. P. Wang, Y. Wang, J. Xu, Y. Cao, X. Wang, X. Cheng, G. Li, and X. Li, Facile metagrating holograms with broadband and extreme angle tolerance, *Light Sci. Appl.*, 7(1):78, 2018. DOI: 10.1038/s41377-018-0075-0. 56

[21] B. Shen, P. Wang, R. Polson, and R. Menon, An integrated-nanophotonics polarization beamsplitter with $2.4 \times 2.4 \ \mu m^2$ footprint, *Nat. Photon.*, 9(6):378–382, 2015. DOI: 10.1038/nphoton.2015.80. 57

[22] Y. Zhang, Z. Li, W. Liu, Z. Li, H. Cheng, S. Chen, and J. Tian, Spin-selective and wavelength-selective demultiplexing based on waveguide-integrated all-dielectric metasurfaces, *Adv. Opt. Mater.*, 7(6):1801273, 2019. DOI: 10.1002/adom.201801273.

[23] Z. Xie, T. Lei, H. Qiu, Z. Zhang, H. Wang, and X. Yuan, Broadband on-chip photonic spin hall element via inverse design, *Photon. Res.*, 8(2):121–126, 2020. DOI: 10.1364/prj.8.000121.

[24] Z. Xie, T. Lei, F. Li, H. Qiu, Z. Zhang, H. Wang, C. Min, L. Du, Z. Li, and X. Yuan, Ultra-broadband on-chip twisted light emitter for optical communications, *Light Sci. Appl.*, 7(4):18001, 2018. DOI: 10.1038/lsa.2018.1. 57

CHAPTER 7

Summary and Outlook

Holography was an old technology but was persistently studied for more than 70 years in the optics community due to its powerful ability to arbitrarily shape free-space wavefront by controlling "holo" attributes typically including both amplitude and phase of the light field. Although many fascinating applications such as holographic data storage, real 3D image display have been demonstrated based on traditional holography, there were still tough problems including low diffraction efficiency, narrow viewing angle and cumbersome optical setup for both recording and reconstruction process.

The merging of holography and nanophotonics brings about a number of opportunities for versatile molding of both far-field wavefront and near-field propagations of light. SPP waves were first introduced for holography technology, assisting to enhance the diffraction efficiency [1, 2] or construct the white-light holography for full-color 3D image display [3]. Then, metamaterials with spatially varying effective refractive indexes were considered to construct holography with small pixel sizes [4, 5]. Finally, single-layered metasurfaces with abrupt phase discontinuity were demonstrated as an excellent platform for high performance and multifunctional holography [6].

Fundamentally different from the traditional phase modulation that relies on the accumulated light propagation inside a thick medium, metasurface holography exploits the scattering phase of the subwavelength meta-atoms by varying its geometric parameters along the metasurface plane. The scattering phase of the engineered meta-atoms includes the resonant phase, geometric phase, propagation phase, and detour phase. Metasurfaces based on different phase modulations have different characteristics. For example, resonant phase and propagation phase metasurfaces have strong dispersion with respect to wavelength, while geometric phase and detour phase metasurfaces have dispersionless feature for phase modulation and therefore work ideally for broadband. Different types of phase modulation rules can be combined to design metasurfaces with extended functionalities. Geometric phase and propagation phase can be combined to independently control the phase and polarization states of the wavefront [7, 8]. Geometric phase and resonant phase can be combined to simultaneously modulate the phase profile and phase dispersion, leading to broadband achromatic metalens [9, 10]. Geometric phase and resonant spectra of meta-atoms can be combined to simultaneously modulate phase and amplitude [11], or phase and frequency of light [12]. Geometric P-B phase and detour phase can also be combined to simultaneously control phase, amplitude, and polarization states of light in a broadband and wide-angle range [13].

Due to large number of degrees of freedom of the meta-atom, metasurface holography is capable to integrate multiple holographic information inside a single layer. Typical ways to increase the information capacity includes the polarization multiplexing and color multiplexing.

For the polarization multiplexing, dual holographic information can be carried by two orthogonal polarization states. By designing anisotropic meta-atoms in resonant phase, propagation phase, or detour phase metasurfaces, the multiplexed dual holography can be imposed on two orthogonal linear polarizations. Circular polarization multiplexing can be realized in geometric P-B phase metasurfaces, as the geometric P-B phase is dependent on the helicity of the circular polarized incident light. Independent dual holographic images can also be imposed on two arbitrary elliptical polarization states with the combined geometric phase and propagation phase metasurfaces, leading to fascinating applications including arbitrary spin to orbital conversion, spin-controlled arbitrary accelerating beams, and metasurface quantum entanglement. Metasurface even provides a powerful platform that can generate vectorial holography with arbitrary polarization state distribution in a holographic image, beyond the dual holographic information multiplexing.

For color multiplexing in metasurfaces, interleaving and non-interleaving approaches were typically applied to realize multiple holographic information encoding by multiple wavelengths. Multi-color metasurface holography was first demonstrated on interleaved resonant metasurfaces, in which multiple types of meta-atoms for multiple wavelength response were interleaved in a super unit cell. Since both the spectral response and phase modulation were dependent on the resonance properties of their meta-atoms, their modulated phase levels were limited and diffraction efficiency and image fidelity were far from satisfactory. Later on, interleaved geometric metasurfaces were proposed to achieve full-color image construction. Since the spectral response and phase modulation were completely decoupled, the phase modulation levels can be largely increased and thus both the efficiency and quality of the full-color holographic image were improved. Non-interleaving approach was another efficient way to realized full-color holography with only single type of meta-atoms. In the recording process, certain phase-shift terms were added for holograms of different wavelength; in the reconstruction process, different colored light beams were set with different proper incident angles to build the predesigned full-color holographic image. With multiple degrees of freedom of meta-atom to be controlled, color multiplexing can be combined with the polarization multiplexing to further increase the multiplexing channels of metasurface holography.

In addition to the typical way to construct metasurface holography by discrete phase gradient meta-atoms, metagratings that connect the surface wave and propagating wave through momentum exchange between diffraction orders reveal another powerful platform for both far-field and near field holography. The metagrating implemented in a plasmonic film etched with holographic positioned subwavelength slit array can convert free-space light beams to arbitrary shaped wavefront of surface waves. Similar metal groove metagratings that follow the interference pattern between a surface reference beam and a free-space object wave can convert surface waves to free-space holographic images. The localized and propagating surface plasmon mode in the metagrating can also perfectly route light beams between desired diffraction orders, providing a convenient way to continuous modulate the phase profile of free-space wavefront in a broadband and wide-angle range.

Until now, metasurface holography have matured in fundamental aspect and abundant state-of-art applications including 3D holographic display, encryption, and digital encoding. Although great progresses have been made, a few of challenges still need to be conquered before the transformation from state-of-art metasurface devices to industrial products in the future.

Nowadays, the metasurface holograms are mainly fabricated on high-cost and time-consuming nano lithography technologies such as electron beam lithography and focused ion beam etching, hindering the possibility for mass production of metasurface holographic devices. In the future, development of alternative metasurface design strategy for simultaneous simplified structure and high performance is needed, so that low-cost nano-imprinting can massively fabricate those devices.

On the other hand, the dynamic wavefront shaping of light is eagerly need especially for the real 3D displays in nowadays television and smart phones. In traditional holography, the dynamic wavefront shaping can be realized in SLM, which is composed of the electrical tunable liquid crystals. However, it is difficult for metasurface hologram to be dynamical tunable because the phase modulation is mainly based on the geometric shape of meta-atoms, which are fixed once fabricated. There were other ways to construct reconfigurable metasurfaces by dynamically tunning material properties of the meta-atom. For example, in terahertz regime, graphene [14, 15] is a promising tunable material by either chemical or electric way; and in microwave regime, electrically tunable printed-circuit-board [16], liquid-metal [17] can be used to dynamically control the holographic images. In optical regime, phase change material [18–20], conducting oxide [21] can modulate the phase profile in a reconfigurable way, but the phase profile are changed as a whole. Very recently, dynamic control of local phase was recently demonstrated on an electric tunable dielectric metasurfaces, functioning like a metasurface version of SLM. However, it can only perform simple wavefront steering applications, far from the arbitrary holographic wavefront shaping dynamically [22]. As a result, the dynamic point-by-point phase modulation is highly desired through ectro-optic or all-optical control, to ultimately realize the real-time dynamic 3D holographic display in an integrated nano structure.

7.1   REFERENCES

[1] S. Maruo, O. Nakamura, and S. Kawata, Evanescent-wave holography by use of surface-plasmon resonance, *Appl. Opt.*, 36(11):2343–2346, 1997. DOI: 10.1364/ao.36.002343. 61

[2] G. P. Wang, T. Sugiura, and S. Kawata, Holography with surface-plasmon-coupled waveguide modes, *Appl. Opt.*, 40(22):3649–3653, 2001. DOI: 10.1364/ao.40.003649. 61

[3] M. Ozaki, J.-i. Kato, and S. Kawata, Surface-plasmon holography with white-light illumination, *Science*, 332(6026):218–220, 2011. DOI: 10.1126/science.1201045. 61

[4] B. Walther, C. Helgert, C. Rockstuhl, F. Setzpfandt, F. Eilenberger, E.-B. Kley, F. Lederer, A. Tünnermann, and T. Pertsch, Spatial and spectral light shaping with metamaterials, *Adv. Mater.*, 24(47):6300–6304, 2012. DOI: 10.1002/adma.201202540. 61

[5] S. Larouche, Y.-J. Tsai, T. Tyler, N. M. Jokerst, and D. R. Smith, Infrared metamaterial phase holograms, *Nat. Mater.*, 11(5):450–454, 2012. DOI: 10.1038/nmat3278. 61

[6] Z.-L. Deng and G. Li, Metasurface optical holography, *Mater. Today Phys.*, 3:16–32, 2017. DOI: 10.1016/j.mtphys.2017.11.001. 61

[7] A. Arbabi, Y. Horie, M. Bagheri, and A. Faraon, Dielectric metasurfaces for complete control of phase and polarization with subwavelength spatial resolution and high transmission, *Nat. Nanotechnol.*, 10(11):937–943, 2015. DOI: 10.1038/nnano.2015.186. 61

[8] J. P. Balthasar Mueller, N. A. Rubin, R. C. Devlin, B. Groever, and F. Capasso, Metasurface polarization optics: Independent phase control of arbitrary orthogonal states of polarization, *Phys. Rev. Lett.*, 118(11):113901, 2017. DOI: 10.1103/physrevlett.118.113901. 61

[9] W. T. Chen, A. Y. Zhu, V. Sanjeev, M. Khorasaninejad, Z. Shi, E. Lee, and F. Capasso, A broadband achromatic metalens for focusing and imaging in the visible, *Nat. Nanotechnol.*, 13(3):220–226, 2018. DOI: 10.1038/s41565-017-0034-6. 61

[10] S. Wang, P. C. Wu, V.-C. Su, Y.-C. Lai, M.-K. Chen, H. Y. Kuo, B. H. Chen, Y. H. Chen, T.-T. Huang, J.-H. Wang, R.-M. Lin, C.-H. Kuan, T. Li, Z. Wang, S. Zhu, and D. P. Tsai, A broadband achromatic metalens in the visible, *Nat. Nanotechnol.*, 13(3):227–232, 2018. DOI: 10.1038/s41565-017-0052-4. 61

[11] L. Liu, X. Zhang, M. Kenney, X. Su, N. Xu, C. Ouyang, Y. Shi, J. Han, W. Zhang, and S. Zhang, Broadband metasurfaces with simultaneous control of phase and amplitude, *Adv. Mater.*, 26(29):5031–5036, 2014. DOI: 10.1002/adma.201401484. 61

[12] B. Wang, F. Dong, Q.-T. Li, D. Yang, C. Sun, J. Chen, Z. Song, L. Xu, W. Chu, Y.-F. Xiao, Q. Gong, and Y. Li, Visible-frequency dielectric metasurfaces for multiwavelength achromatic and highly dispersive holograms, *Nano Lett.*, 16(8):5235–5240, 2016. DOI: 10.1021/acs.nanolett.6b02326. 61

[13] Z.-L. Deng, M. Jin, X. Ye, S. Wang, T. Shi, J. Deng, N. Mao, Y. Cao, B.-O. Guan, A. Alù, G. Li, and X. Li, Full-color complex-amplitude vectorial holograms based on multi-freedom metasurfaces, *ArXiv:1912.11184*, 2019. DOI: 10.1002/adfm.201910610. 61

[14] Z. Miao, Q. Wu, X. Li, Q. He, K. Ding, Z. An, Y. Zhang, and L. Zhou, Widely tunable terahertz phase modulation with gate-controlled graphene metasurfaces, *Phys. Rev. X*, 5(4):041027, 2015. DOI: 10.1103/physrevx.5.041027. 63

[15] P. C. Wu, N. Papasimakis, and D. P. Tsai, Self-affine graphene metasurfaces for tunable broadband absorption, *Phys. Rev. Appl.*, 6(4):044019, 2016. DOI: 10.1103/physrevapplied.6.044019. 63

[16] L. Li, T. Jun Cui, W. Ji, S. Liu, J. Ding, X. Wan, Y. Bo Li, M. Jiang, C.-W. Qiu, and S. Zhang, Electromagnetic reprogrammable coding-metasurface holograms, *Nat. Commun.*, 8(1):197, 2017. DOI: 10.1038/s41467-017-00164-9. 63

[17] P. C. Wu, W. Zhu, Z. X. Shen, P. H. J. Chong, W. Ser, D. P. Tsai, and A.-Q. Liu, Broadband wide-angle multifunctional polarization converter via liquid-metal-based metasurface, *Adv. Opt. Mater.*, 5(7):1600938, 2017. DOI: 10.1002/adom.201600938. 63

[18] Q. Wang, E. T. F. Rogers, B. Gholipour, C.-M. Wang, G. Yuan, J. Teng, and N. I. Zheludev, Optically reconfigurable metasurfaces and photonic devices based on phase change materials, *Nat. Photon.*, 10(1):60–65, 2016. DOI: 10.1038/nphoton.2015.247. 63

[19] A. Karvounis, B. Gholipour, K. F. MacDonald, and N. I. Zheludev, All-dielectric phase-change reconfigurable metasurface, *Appl. Phys. Lett.*, 109(5):051103, 2016. DOI: 10.1063/1.4959272.

[20] C. H. Chu, M. L. Tseng, J. Chen, P. C. Wu, Y.-H. Chen, H.-C. Wang, T.-Y. Chen, W. T. Hsieh, H. J. Wu, G. Sun, and D. P. Tsai, Active dielectric metasurface based on phase-change medium, *Laser Photon. Rev.*, 10(6):986–994, 2016. DOI: 10.1002/lpor.201670068. 63

[21] Y.-W. Huang, H. W. H. Lee, R. Sokhoyan, R. A. Pala, K. Thyagarajan, S. Han, D. P. Tsai, and H. A. Atwater, Gate-tunable conducting oxide metasurfaces, *Nano Lett.*, 16(9):5319–5325, 2016. DOI: 10.1021/acs.nanolett.6b00555. 63

[22] S.-Q. Li, X. Xu, R. M. Veetil, V. Valuckas, R. Paniagua-Domínguez, and A. I. Kuznetsov, Phase-only transmissive spatial light modulator based on tunable dielectric metasurface, *ArXiv:1901.07742*, 2019. DOI: 10.1126/science.aaw6747. 63

Authors' Biographies

ZI-LAN DENG

Zi-Lan Deng is an Associate Professor at the Institute of Photonics Technology in Jinan University, China. He obtained his Bachelor's and Ph.D. degrees from Sun Yat-Sen University, China in 2009 and 2014, respectively. From 2014–2016, he held postdoctoral positions at College of Electronic Science and Technology (Institute of Microscale Optoelectronics), Shenzhen University, China. After that, he joined the Institute of Photonics Technology, Jinan University. From 2019–2020, he was working as a visiting scholar at Advanced Science Research Center, City University of New York. His research is mainly focused on surface plasmon waves, vectorial metasurfaces, and holography.

XIANGPING LI

Xiangping Li is currently a principle investigator at the Institute of Photonics Technology in Jinan University, China. He obtained his Bachelor's and Master's degrees from Nankai University in 2002 and 2005, respectively. In 2009, he completed his Ph.D. at Centre for Micro-Photonics in Swinburne University of Technology. From there he joined the Centre for Micro-Photonics as a postdoctoral research staff, and served as a senior research fellow between 2013 and 2015. From 2015, he joined Jinan University and founded the Nanophotonic Device Group. His research is focused on nanophotonics including metasurface, optical multiplexing, and optical data storage.

GUIXIN LI

Guixin Li is currently an Associate Professor, Department of Materials Science and Engineering, Southern University of Science and Technology, China. Before this position, he was a Research Fellow at Department of Physics, University of Paderborn and the Metamaterial Research Centre, School of Physics and Astronomy, the University of Birmingham from 2014–2016. He was a Research Assistant Professor in the Department of Physics, Hong Kong Baptist University between 2012 and 2014. Dr. Li obtained Ph.D. from Physics Department, Hong Kong Baptist University in 2009 and Bachelor's degree in Physics, Beijing Normal University in 2003. Guixin Li has published more than 70 peer-review papers in high impact journals such as *Nature Materials, Nature Nanotechnology, Nature Physics, Nature Reviews Materials, Nature Communications, Physical Review Letters,* and so on. Google citations exceed 4500. Dr. Li was awarded the 2019 Qiushi Outstanding Young Scholar, China.